The Ethical Digital Transformation

The Ethical Digital Transformation

Alain Onesti

COPYRIGHT © 2022 (Alain Onesti)

All Rights Reserved

ISBN-13

To my family,
for always supporting me over the years

Contents

Introduction 1-2

1. Those In Charge of Implementing Digital Change 1-7

    Becoming A Manager of Digital Transformation 1-8

    The Role of Leadership 1-9

    Taking The Initiative to Bring About Change 1-13

    Is It Possible for Managers to Lead A Digital Transformation If They Lack The Necessary Skills, Capabilities, And Courage? 1-18

2. Working People Affected by The Digital Transformation 2-21

    As A Result of Advances In Digital Technology 2-22

    New Paradigms? 2-25

    Digitalization 2-30

    A Shift in The Nature of Work 2-35

3. Innovating New Roles for Workers Who Undergo The Digital Transformation Process 3-38

    Is It Feasible to Reimagine The Workplace? 3-39

    New Technology Can Be Put to Use in The Real World 3-44

    The Digital Welfare State's Function 3-45

    Digital Upheaval of The Established Order 3-46

    Where To Go From here? 3-47

4. Change Is Inevitable in The Age of Digital Transformation, But It May Also Be an Opportunity to Stretch One's Boundaries 4-50

    What Is Your Company's Greatest Opportunity and Worst Threat? 4-51

    Connect With Organizations & Societies 4-53

Opportunity To Invest in The Stock Market..........4-56

Models of Digital Businesses 4-58

Is It Worthwhile? 4-62

5. Digital Transformation Is a Friend Or A Threat?5-64

The Shadowy Side of Things 5-66

Risks to Security 5-67

Infrastructures Utilizing a Combination Of Public And Private Cloud Services 5-69

Supply Chains and Channels for Sales In The Digital Age.................5-70

The Interconnectedness of Everything (IoT).........5-72

Automated And Data-Driven Decision Making.....5-73

Customer Satisfaction Is Improved 5-74

Agility Is Boosted.................5-75

Benefits for Industries.................5-76

Enhanced Data Accessibility 5-77

Improved Workplace Culture.................5-77

Foreshadowing Of What Is to Come 5-78

6. Transfer of Data 6-79

Cleansing The Data.................6-80

Stations For Storing Personal Data 6-83

Preparation of Data.................6-83

Encryption Of Sensitive Information 6-86

Protecting The Confidentiality of Personal Information6-88

Software for Data Transfer 6-90

Look into the Source Codes 6-92

Carry out the Transfer.................6-93

New System Transition 6-93

Ongoing Improvements should be managed and monitored 6-93

7. Interconnection of various systems (old and new) the API notion... 7-95

    What does API Integration mean?........................... 7-96

    Why Do You Need? 7-98

    Methods for Integrating API 7-100

    Integrate existing systems into one 7-101

    An example of a common API 7-103

    Real time, current and efficient results can be achieved by using this tool 7-104

    APIs: The Good, the Bad, and the Dubious 7-105

    Getting the Right Amount of Both 7-106

    API Integration Has Many Advantages................ 7-107

8. Importance of training for business users and for the adoption of new system.. 8-109

    Development of a New Technology Training Program Is Crucial 8-110

    Avenues to take a lot... 8-111

    Knowledge of Digital Marketing Can Help You Improve Your Customer Experience 8-113

    Training in digital marketing teaches you how to make the most of your digital channel 8-113

    Investing in the digital skills of your employees improves both communication and productivity 8-114

    Definition of user adoption................................... 8-115

    What is the significance of user adoption 8-115

    Understanding how the product will be used is critical 8-116

    The foundation of a user adoption strategy is training 8-117

    Maintain a diverse training program.................... 8-117

Training pays off in the long run 8-118

9. Designing a solution and migrating to the cloud ... 9-119

Development in a coding language that does not require a lot of 9-120

A Quicker Changeover 9-121

Interactiveness 9-122

Cost Reduction 9-122

Advanced Personnel Skills 9-123

What is cloud migration, and how does it differ from traditional migration? 9-123

Process of moving to the cloud 9-124

The advantages of moving to the cloud 9-126

Security 9-127

Scalability 9-128

Price Factor 9-129

Integration 9-131

Access 9-132

10. Obligatory Role of Ethics 10-134

Digital businesses must adhere to business ethics 10-136

Design for confidentiality, integrity, and safety . 10-137

Inspire faith in others 10-138

Be wary of preconceived notions 10-140

Ensure that there is accountability 10-141

Promoting a more ethical way of life 10-141

Digital Success 10-142

Ethical Career Development 10-143

Think ethically about technology 10-144

Foreword

Alain Onesti is an IT professional with twenty years of experience in the world of digital transformation.
He has worked for large companies and well-known multinationals managing digital transformation projects that involved different sectors and different nations.
During his experience he paid particular attention to the ethical aspect by developing win-win paths for both the companies and the employees involved.

Introduction

As customer expectations continue to rise, so does the requirement for enterprises to adapt to market dynamics with ever-increasing levels of business agility through digital transformation. Cloud computing, robotics and AI, big data and efficient operational models enable firms to innovate and adapt to internal and external events faster and more cost-effectively than ever before. These observations should serve as a reminder that the human factor is just as crucial as technology in enabling digital transformation. In order to achieve long-term success in a digital environment, organizations must first and foremost recognize the importance of acting ethically, which is a core tenet of ethics.

Customers' expectations for products and services have changed, resulting in a reassessment of how an organization leverages technology, people, and processes to pursue new business models and new revenue sources. The development of digital products, such as mobile applications or an e-commerce platform, has become a priority for many traditional goods-producing companies. Cross-departmental collaboration is critical to the success of digital transformation, which should be led by CEOs and other senior executives in tandem with CIOs, CHROs, and other IT leaders.

Even though businesses have been attempting digital transformation for years to stay ahead of both incumbents and startups, progress has been slow – until the outbreak of the epidemic. With the introduction of curbside delivery and other contactless payment alternatives, retailers such as Walmart and Bed Bath & Beyond revolutionized their shop operations. Cloud-based video collaboration software and machine learning (ML) software for supply chain management have been reprioritized in strategic IT roadmaps by IT leaders, with many selecting the latter. When implemented point by point, they don't facilitate change. When these tools and other solutions are integrated into a company's workflow, it is easier to gauge its digital health and reflect its business goals.

Rather than focusing on what people actually do, ethics focuses on what they should do in order to make moral decisions. In contrast to the past, when the subject was not widely accepted by the corporate world, things have changed dramatically. For a highly competitive market where a company's reputation and values are as crucial as its products and services, ethics is no longer just something businesses care about; it has become a differentiator. While ethics is certainly something that all firms with a digital transformation goal should aggressively embrace, the most challenging task will be at the individual level — corporations do not make decisions, individuals do.. Ethics is no longer only a matter of "right" vs. "wrong"; depending on your point of view, there may be more than one "correct" response. It is not uncommon for people to be faced with ethical dilemmas in which they must choose between two or more unpalatable options, each of which has implications for other parties that must be taken into account.

However, little research has been done on the role of ethics in the digital transformation. There are various ethical issues raised by the increasing use of digital technologies and the shift to digital business models. When speed takes precedence over other considerations, the perceived performance of ethical behavior may be badly impacted. An automated decision-making process could also put customers' rights to fair treatment and their ability to make creative choices in jeopardy. Data gathering and storage made possible by new technology may also put customers' personal information at risk of being misused. It is possible that companies sharing customer data among or with partners will exacerbate this issue.

The 'correct' thing to do from an ethical perspective will have to be determined by digital professionals and executives. Regardless of an organization's level of ethical awareness or ethics training, what is considered ethical might differ between individuals, groups, religions, and cultures, and these allow substantial opportunity for interpretation in a global and fast-moving digital society. In the real world, even if the appropriate path of action is evident, people are often compelled to make judgments that could harm others. As a result, being ethical means having the ability and moral bravery to defy conventional wisdom and act ethically.

Publishers must adopt new technology in order to enable the shift away from printed material to digital content. Apps and websites can be used to distribute digital information, while artificial intelligence can be used to supply clients with individualized content. Ethical concerns have been raised about a number of these emerging technologies. Two major ethical issues arise in the context of emerging technologies. As new tools can be used in multiple ways, businesses must first choose the best use for each one. It's a question publishers must grapple with as well: Do recommender systems limit customers' ability to be creative?

In both the now and the future, digitization is the driving force behind major economic and social developments in our society. Artificial intelligence (AI), blockchain, and other cutting-edge (data) technologies have huge societal and economic possibilities. When it came to dealing with the COVID-19 situation, for example, we witnessed businesses and governments unquestionably relying on digital technologies such as working from home technology, corona apps, and drones. The public debate and the media attention proved that we cannot put our complete faith in digital technologies. And that confidence is critical, because the usage of unknown digital technology is costing society and organizations a lot of money.

It is now more critical than ever for enterprises to pay attention to digital ethics, as we are on the verge of the next wave of digital transformation that has a greater impact than ever before. Developing and implementing digital technology in a responsible manner requires a business to adhere to certain ethical norms. Digital ethics guide an organization's behavior toward its stakeholders in this way.

1. Those In Charge of Implementing Digital Change

The manager of a digital transformation must possess not only strong project management abilities, but also a thorough knowledge of the organizational structures of modern businesses, as well as strong capabilities in managing change. This is a challenging position that necessitates the use of a wide range of abilities. They must be able to communicate and motivate others, as well as comprehend and analyze business processes, in addition to having the requisite technical skills. They need to be able to deal with change, but they also need to be able to recognize it and take advantage of it. Finally, they should have a strong concentration on achieving their goals. Since this is not solely a technical function, the digital transformation manager is often drawn from the ICT department (where they may serve as CTO or CDO). Indeed, the ability to comprehend and understand the latest technical advancements (such as big data, artificial intelligence, and machine learning; the internet of things; and cyber security) is a prerequisite for any firm.

A course can be an option if you're interested in this position but lack the requisite abilities.

Becoming A Manager of Digital Transformation

However, what exactly should a manager in charge of digital transformation do? To begin, a digital transformation manager should have a comprehensive understanding of the organization, including all of its parts and operations (stakeholders, management, employees, clients). They must be aware of the current state of digitalization and the potential opposition they may face, taking into account the complete ecosystem of their firm. Among other things, they should examine the organization's digital efficiency as a whole, as well as the organization's technological and digital infrastructures. Then she needs to identify the goals she wants to achieve and devise a strategy for achieving them. If the goals are little, they don't have to be. They could be as simple as streamlining the lines of communication or as complex as automating time-consuming, repetitive tasks. They might also be the development of artificially intelligent technologies to assist with marketing decisions. Digital transformation managers are also responsible for allocating funds, managing the people involved, and conveying their views to higher ups. It is imperative that she monitor and audit the innovation process throughout its existence. A role that is both critical and exhilarating at a time when everything is constantly shifting.

The Role of Leadership

Managers' responsibilities in businesses must expand. These unseen, unwritten habits that are difficult to assess, but which ultimately shape company culture, have become increasingly commonplace in today's business world. By definition, corporate culture is the sum of the individual actions of the company's managers, which is the outcome of the existing management practices in existence. In new organizational forms (freedom-form corporations, holacracy, startups, etc.), the title "manager" may disappear, although traditional management functions continue to exist. You'll still need to know how to run a project, assign tasks, delegate authority, etc. There has been significant discussion in recent years about the transition from managers to coaches. This change is taking place and demonstrates the problems posed by digital transformation. Managers have an important role to play when half of the company's employees are unable to upload a video in less than one minute. It is the manager's responsibility to not only communicate the strategic vision of the company, but also to give the tools necessary for their staff to grasp this vision and apply it to their job.

Managers play an important role in providing day-to-day assistance. To transform, it's not a matter of skill as much as it is a matter of adapting one's lifestyle to meet the new culture. Allowing employees to explore and learn, and helping them foresee their existing skill set's obsolescence, is the key to empowering them. Managers tend to take on a new role when a company is constantly learning and adapting: centralizing, enabling, and pushing internal best business practices and methodologies. A leader's job isn't done until he or she encourages the members of their team to document and share their best practices. In order for a company's collective intelligence to grow, managers are responsible for developing a work environment that fosters innovation. When a team is going through a generational shift, this job is critical. It is critical that senior staff members and junior employees (and vice versa) work together to pass on their knowledge. This saves time and improves productivity by utilizing the company's own knowledge and making it actionable.

The manager is also responsible for implementing new work practices at the team level. It's common for large-scale digital transformation initiatives to fail because they focus on the entire organization instead of focusing on individual individuals and their micro-practices.

Identifying learning opportunities based on strategic company goals should be the primary responsibility of a manager in order to promote this kind of natural skill enhancement. Self-improvement should be encouraged and supported by managers as well. When training was once done at set times or dates, our relationship with learning has altered since we are always training and learning new things. This also suggests that managers offer their employees time to learn, such as by allowing them to regulate their own schedules and setting aside time specifically for training.

It is the responsibility of managers to help their staff see the value of different points of view and the aims of other divisions within the organization. Workers must comprehend their work as an ecosystem rather than a series of pointless tasks that have no clear, quantifiable results for the employee.

The appropriate leadership is essential for companies undergoing digital transformation. Digital technologies necessitate a shift in management methods. Traditional corporate methods are giving way to new forms of leadership as the nature of leadership itself evolves. Recently, a new post known as Chief Digital Officer (CDO) has emerged. An organization's digital transformation is the responsibility of a company's Chief Digital Officer (CDO). Since the CDO's primary responsibilities revolve around digital transformation, the CIO's primary responsibility is to manage the IT infrastructure of the company. To put it another way, digital transformations are establishing CDOs as new executives at the top management level of companies. The appointment of a Chief Digital Officer (CDO) shows how critical digital transformation has become to the company as a whole. Every department has a digital business strategy that the CDO is charged with communicating and implementing. Rather than focusing on a single department, the CDO tries to restructure and energize the entire organization. It's a multifaceted project that revolves around digital transformation. The first step is to create a strategy for the company's transformation. The CDO should be involved in this process with all of the stakeholders that the transformation touches. Furthermore, the CDO should work to improve collaboration between IT and business units, which means that the digital technologies used should be aligned with the firm's strategic objectives. The CDO is responsible for dealing with resistance during the implementation of the transformation strategy. In addition, the CDO is working to instill a company-wide climate of openness to new ideas and experimentation. This requires that all employees of the company have a digital attitude. In addition, personnel must be prepared to deal with the inevitable interruptions brought on by the adoption of new digital technology. To be successful in digital transformation, however, a CDO post is not enough. The involvement and support of senior management is critical to the transition. As a result, CDOs face one of the most significant challenges: a dearth of top-level support. Resistance from employees can only be overcome if top management is totally on board with a reform process. There should be a clear understanding and belief in its benefits from the top management. Furthermore, the CDO should be given the authority to implement digital transformation by top management. It's possible that the CDO post is only temporary, which means that digital transformation may also reach a conclusion at some point in the future. Ethics and leadership go hand in hand. The ethical culture of a company is heavily reliant on the leadership's ability to uphold high standards of conduct.

Taking The Initiative to Bring About Change

Top executives play an active role in the digital transformation by setting the tone for the transformation. Employees may see the digital transformation as a danger because of the huge changes it represents for firms, but because humans value a sense of coherence, consistency, or continuity through time. Thus, they are more likely to oppose change, which has a negative impact on the success of planned organizational reforms. Top managers have a critical role to play in fostering employee acceptability and, as a result, winning their allegiance. Managers guide their teams through a moment of digital transition by signaling personal commitment," "engaging others," and creating trust and commitment. When it comes to demonstrating their own personal dedication to digitization, the company's senior executives demonstrate that they share their company's belief in its chosen course. Leaders' communication of a change vision, which represents the organization in a possible future state in which the organizational identity endures, is a key medium for mobilizing followers toward support for change, according to leadership study Employee mailings, presentations, and other internal networking opportunities are all ways that managers demonstrate their commitment to the digital transformation of their companies. They openly discuss the topic and highlight lighthouse projects that demonstrate the importance of it. As a role model or pioneer in the field of digital transformation, senior executives demonstrate their dedication to it by pushing the issue despite difficulties or criticism from others. As part of the company's digital transformation, managers make it a point to involve others in the process. Many interviewees emphasized the importance of open and transparent communication because of the significant changes that digital transformation will bring to businesses and their employees. Preliminary studies show that employees' willingness to embrace change increases when their managers communicate regularly and credibly with them in words and actions that support their claims. Efforts are made to pique the interest of employees in digitalization themes and to promote their active engagement in the process. TMT experts and departments are given the duty and funding to design their own solutions, according to one respondent's description. In this setting, ideas and solutions are more likely to succeed. It's not uncommon for senior executives to actively encourage their employees' use of digital tools, and to reward them for their efforts. A variety of events (such as start-up pitches and expert speeches) as well as networking possibilities (such as roundtables and social communities) are some of the ways managers aim to pique the interest of their employees and win their loyalty (e.g., digital learning platforms and workshops).

Managers try to overcome the difficulty of change by fostering a sense of trust and commitment in the workforce. As a result, employees are more likely to accept and support suggested organizational changes if leaders are perceived to be credible in their explanations. Managers build a culture of openness and risk-taking in their companies, encouraging their employees to learn from their mistakes and improve their performance. By reassuring workers that their jobs would still exist in the future, senior managers hope to allay their anxieties about the digital transformation. It's not just that senior managers are aware of this changing reality, but they also try to handle it by openly communicating with the workforce and seeking for a solution that would benefit everyone.

Authenticity is essential for the digital transformation. Change management, not technology, is the most important consideration. Taking individuals by the hand, removing their worries, and persuading them that digitalization will have a profound impact on our working world. But, of course, there will be new opportunities.

Strategic management research has long focused on the importance of top-level management's ability to shape an organization's culture. There has been a significant amount of study based on the upper echelons' idea about the causes and effects of top management on individual firms. Top managers' qualities (such as demographics, personality traits, or values) have been the focus of relevant studies in an attempt to represent their underling cognitive basis. Specifically, senior managers have a considerable impact on the firm's strategy and performance because of their decision-making power and leadership. A formal definition of 'managerial discretion' captures how much senior managers can influence firm-level outcomes. The increasing latitude granted to executives, and in particular to CEOs, has highlighted the necessity of delving deeper into the implications of their decisions and actions. To remain competitive in today's market, CEOs must devise and implement a business strategy that takes into account the potential and dangers of the digital transition. Only seldom have the antecedents and consequences of the characteristics and actions of top managers taken into account aspects of digitalization.

Managers of digital transformation are primarily in charge of directing and designing the company's digital transformation strategy. As a vital aspect of the company, they link teams with determining business needs and searching for technology capabilities within the company's internal infrastructure. In this method, they conduct specific research to determine the IT needs of the workforce. They must guarantee that the right IT infrastructure is purchased and installed on-site. They maintain the organization's digital sovereignty while ensuring the stability of IT procurement and digital change. When purchasing software solutions or services, cost-effectiveness and productivity are taken into consideration. They devise methods for persuading people to make use of newer forms of technology. Managers must ensure that data analytics capabilities are in place while implementing digital business transformation. Security is a major worry when digital transformation occurs. To make sure your systems are safe, have your digital transformation manager look over your cybersecurity practices. For each issue, they investigate the situation and put together a team of experts to help resolve the issue.

Is It Possible for Managers to Lead A Digital Transformation If They Lack The Necessary Skills, Capabilities, And Courage?

Digital transformation necessitates a high level of managerial expertise. It's essential that they have a firm grasp on their own unique set of skills and the way their company makes use of data. Understanding these principles is essential for today's managers, who must also know how to obtain and budget financial resources. Information technology has increasingly intertwined with traditional company tasks including operations, accounting, finance, and sales & marketing. Each and every one of a company's employees is in some way impacted by the company's information systems. Managers that are aware of this do not rely only on their IT departments to make technology-related choices for their companies. Strong leaders are also aware that the informational needs of a company might change across the organizational chart and that different levels of management may require different informational technology. IT departments do not have the authority to make judgments about how and what technology their companies require.

Many failures in business may be traced back to poor digital transformation decisions, but success stories can't always be linked that way. As crucial as a manager's capacity to guide their organization through the ever-changing information systems in their business ecosystems is their ability to communicate effectively with their employees. Digital transformation leaders need to know which tech solutions will give the best service and value for both the company and its customers, since customers increasingly expect a variety of digital experiences, from online buying to video conference sales meetings to simple smartphone support interactions. Having a strong interest in technology is a must, since this will keep them abreast of the most recent developments and advancements.

- Able to weigh a variety of variables and find a middle ground

Many factors must be considered before the implementation of a new software platform. Digital transformation leaders know how to analyze the value of each platform and its potential usefulness, and then assist the team in learning to use it. There are a number of factors to consider, like the cost, features and security of the product, before they can discover a solution that works for everyone.

- Has the ability to decipher info.

Leadership in digital transformation requires a thorough understanding of client digital experiences in order to pinpoint potential areas for improvement. To obtain a better understanding of their client base and sales strategy, most sales leaders spend a lot of time studying sales reports and other CRM statistics.

- Able to articulate demands.

The traditional method of doing things gets disrupted by digital transformation, and this can generate a lot of stress. You need digital transformation leaders who can communicate the new direction your firm is taking, explain organizational and procedural changes and collect input from employees on what's going well and not so well in the new environment.

- Inside and out knowledge of the company

It's critical to keep an eye on your organization's pulse while leading a digital transition. There's no guarantee that every new tech product that comes onto the market is going to transform your business. A digital transformation leader, on the other hand, can be ready to seize a new digital opportunity as soon as it surfaces by closely monitoring the company's exact demands and operations.

2. Working People Affected by The Digital Transformation

When discussing digital transformation, many argue that computers will eventually replace human labor. As a result, digital transformation might potentially provide new career prospects. Employees don't necessarily think their jobs are under threat from technological advancement. According to a lot of individuals, robotics and artificial intelligence could result in job losses. Most employees, on the other hand, do not believe that computers will ever take the place of humans in the workplace. As a result, the relationship between technical advancement and job insecurity is unclear. Individual well-being can be negatively impacted by work uncertainty and loss, which in turn contributes to social isolation, whereas steady employment gives economic and social benefits.

Increasing attention has been paid to the subjective sense of job insecurity in recent years. Employees' perceptions of their personal status have a big role in the consequences of job insecurity. Uncertainty about one's job has a negative impact on a person's overall well-being, including their physical and mental health. Not only do individual, business, and job qualities affect how individuals feel about their job security, but so do social, political, and macroeconomic contexts as well. As the digital transformation progresses, it is being cited as a factor in perceived job insecurity. Only a few studies have examined the link between technological advancements and workers' feelings of insecurity in the workplace.

As A Result of Advances In Digital Technology

It is the most thorough information revolution of all that is taking place in the digital and electronic media of today. For a long time, we thought that the computer's primary function was to think, to create an artificial intellect that would surpass our own. When a computer named Big Blue defeated chess expert Garry Kasparov, many believed that this goal was within reach. Today, we can clearly see that technology was shifting in a new direction, toward the use of networks to communicate. Increasingly powerful and fast computers are allowing mathematicians and other academics to undertake calculations and simulations that were previously impossible. This is a huge benefit to mathematicians and other researchers. The sum total of our knowledge is increasing at an astronomical rate. But the most intriguing feature of this growth is the worldwide, digital network. The emergence of a new dominating media technology heralds the dawn of a new era. With a relatively minimal investment in technological equipment, and a few easy acts, anyone may become a producer or consumer of both written and visual content on the internet. A more democratic platform than the internet can be hard to imagine; we are all authors and producers, our freedom of speech is as good as total, and our potential audience is incalculably wide open. At the press of a button, you may access vast swaths of material on any subject. Unprecedented expansion has been experienced by this new media. American defense agencies in the 1960s decided to employ computerized networks to decentralize their resources via a network of remote but connected terminals, laying the groundwork for the internet. A nuclear war with the Soviet Union was avoided and the consequences were minimized because to this plan. After the technology proved incredibly effective in organizing joint research initiatives, American and foreign colleges were eventually connected to it. This evolution explains why CERN, the European institute for particle physics in Switzerland, developed the World Wide Web, the technology that ultimately became the internet standard for homepages.

Until the late 1980s, the internet was primarily a tool for the US Department of Defense and the scientific community, and it wasn't until the personal computer and modem breakthroughs in the late 1980s that the internet was changed from ARPANET into public property. Even in the early 1990s, few people were aware of the internet's existence. In December 1995, Bill Gates woke up and said in a memo that Microsoft would be shifting its strategy and focusing on net traffic, one month after an earlier memo stated that Microsoft had no interest in the internet.

The internet has grown tremendously since then. Because the Internet is evolving at such a breakneck pace, it's pointless to provide any statistics on how many computers are currently connected to it. By the time this is read, the data that was accurate when it was written will be completely outdated. This development has elicited a wide range of reactions. Claims of IT-revolutions and new economies are being blown out of proportion, according to critics. Many of these doubters point out that, despite the fact that IT-related stock prices are surging on trend-sensitive stock exchanges around the world, the vast majority of these companies are losing money on a regular basis. There are just a few people who have made a fortune from computers and IT, mostly consultants and the makers of the machines and software that make the internet possible. The economy as a whole has failed to show any signs of exponential development in recent years.

New Paradigms?

The sceptic sees the world as it always was: the same old, same old. However, despite the fact that we still produce hammers, nails, banking institutions are still in business, the impact of these changes has been exaggerated. Instead of utilizing a Dictaphone or a secretary, most people now compose their own business letters using a word processor. However, whether this has improved things significantly remains to be seen. Even if we're employing snazzy new machines, e-commerce is basically business as usual. Following the current trends is more important than being the first to implement them, according to this idea. It doesn't matter whether or not these innovations have any real value. No matter how we communicate, the most essential thing is the substance, not the medium. In the future, the tried-and-true truths of the past will remain such. The opposing viewpoint is overjoyed. Everyone who has seen the light on their screen believes that everything will be OK. All of our issues can be solved through the internet: Ethnic and cultural tensions will go away, replaced by a worldwide digital brotherhood, and the economy will flourish for everyone. All of our tasks as citizens will become more significant as a result, and the democratic system will be reinvigorated in the process. We'll discover the social cohesion we so often lack in today's society on the digital networks, and harmony will permeate across society. Entertainment will become more engaging and entertaining than ever before thanks to this new technology. There is no middle ground between the sceptics and believers. Ignoring the rapid pace of change in which we find ourselves is neither wise nor effective. Both of these viewpoints reveal a reluctance to engage in critical thinking and a failure to see. Prejudices, not analyses or forecasts. The foundations of society, the economy, and culture will definitely be altered by a new, revolutionary communication and information technology. Our difficulties are far from solved, however. To think otherwise is to be nave. While progress allows us to take a more radical approach to some problems, it also brings with it a slew of new issues to contend with. We have the power to extend our lives, improve our health, feel more liberated, and come closer to the goals we set for ourselves. However, the underlying tensions between classes and groupings of people will not disappear, but rather become more complex and difficult to understand. This type of change does not happen instantly. While it is true that we are still in the early stages of the digital transformation, the sceptic who points out that the majority of the global economy is still based on the production of physical goods like refrigerators, planes, and garden furniture rather than digital services on the internet is both impatient and incapable of grasping the full scope of change. The fridge will not go away; rather, the things in our environment will take on new meanings and roles as part of a completely new socio-ecological system. For example, refrigerator advertising will no longer focus on the device's ability to keep milk fresh, as we take that for granted, but on its ability to communicate intelligently over a network. Changes take time to take effect since that's just how things work. The actual hues of any new invention may only be seen after a necessary period of incubation. More than three hundred years after it was first invented, the printing press reached its final breakthrough, causing a major shakeup in social structures and creating a new economic paradigm: capitalism. It took a long time for print to have a positive impact on huge social groupings because reading wasn't ubiquitous enough at the time. There were no indicators of emerging industrialism until the Enlightenment in the 1700s when thinking, information exchange, and technological advancements all reached fever pitch.

It's impossible to meet the needs of a growing population without expansion. Inventions are born out of necessity. There are two ways to look at this: from a purely economic standpoint, an increase in population provides a larger market and a human base for a variety of trials with different types of products that can be tested. A third result is that the economic environment is subjected to two distinct but at least partially conflicting demands: it must be stable while also undergoing some sort of breakdown. For scientific experimentation to be successful, it needs a solid foundation, but on the other hand, there must be an openness to new ideas and the ability to adapt quickly to new developments. Because of machines, mankind's physical power was amplified numerous times over. 'The Industrial Revolution' With the advent of "The Digital Revolution," the human brain will be able to perform hitherto unimaginable feats of computation thanks to electronic networks. But we haven't arrived yet; the prerequisites haven't been met. Even though technology is advancing at an incredible rate, we humans are a step behind. In the face of religious and ideological prohibitions, we are once again stymied. We are once again on the verge of a time of creative devastation that is absolutely essential. There is no way to stop this from happening. For better or for worse, every new technology worth its name has done its own thing, regardless of what its originators had envisioned.

We must learn to live without that luxury and accept that only change is everlasting. Everything changes constantly. The goal and norm of social and economic stability is becoming increasingly rare and a symptom of stagnation.

Rethinking and erasing old thoughts are no longer adequate; they must be constantly rethought. Creative destruction is a never-ending process. An anomaly or crisis is discussed in scientific ideas, where the concept of paradigms was originally formed. An anomaly is a phenomenon that is both unexpected and difficult to fit into the prevailing paradigm. Anomalies Today, they are everywhere: in society, in our cultural and media environments, as well as within our economic systems. At a dizzying pace, political preconditions are changing. The ideological maps of yesterday have nothing to do with today's reality. Branches and empires of the media are coming apart in front of our very eyes. All of our notions about secure employment, automatic promotion, and hierarchical structure are being dismantled by a tremendous upheaval in the workplace. In a matter of months, young men and women who are still wet behind the ears and dressed in unusual clothing are becoming multi-millionaires in firms that few of their stockholders understand.

One of the two possible outcomes is that there are a big number of abnormalities. In the beginning, they tried to fit the new phenomena into the old framework of theories.
An entirely new system of values is taking shape in society, culture, and economy. Electronic networks are expanding at an incredible rate thanks to digitization and other related changes in information management. We're experiencing a

fundamental shift in our mental ecosystem that requires a whole host of other changes as a result. Rather of being set in stone, the new paradigm will be ever-changing. It's not just that we're creating new social standards; it's an entirely different kind of norm.

Digitalization

IoT, big data, mobile apps, augmented reality, social media, and many more digitalization concepts are all included under one umbrella term. The term "digitalization" is commonly used in the business world to describe the process of utilizing digital technologies and digitized resources to improve or change business models and processes in order to generate new sources of value. Many people use the term "digitization" as a synonym for digitization or digital transformation, which is a mistake. But it is important to distinguish between the three. Only the process of converting analogue data (such as images, sounds, etc.) to a digital format (binary code) will be referred to as "digitalization." We will use the term "digitalization" to refer to the business process that was previously discussed. A corporation's transformation into a digital company and the broader ramifications of digitalization on society are both described as "digital transformation" in this context. Mechanization, automation, industrialization, and robotization can also be confused with digitalization. The phrase "digitalization" refers to the production of new sources of value, rather than the improvement of existing processes and workflows.

A task that was previously completed by a human being or was impossible to complete by a human being is referred to be "automated." Machines are being used to take the place of humans in this industry, and this is a closely related field to mechanization. Unlike systems that operate autonomously, which are able to accomplish a goal without the intervention of people, this type of system is able to accomplish a certain aim. In other words, the term "automation" implies that the system follows a predetermined set of rules in order to achieve its aim. Most automated systems are composed of three components: power supplies, feedback control systems, and software for the machines themselves. In this category are the programs and directives that establish the system's desired output and the necessary processes for achieving this goal. This type of automation is referred to by various names, including industrial automation, fixed automation, programmable automation, and flexible automation. Automatic and instantaneous changes can be made to the system thanks to a central computer. Because of this, the system is able to handle multiple tasks at once. Automated systems in the modern period are made up of a variety of technologies.

It's possible that artificial intelligence (AI) could have a significant impact on our lives in the future. It's impossible to pin down exactly what it is because it's so wide-ranging. For tasks that would ordinarily need human judgement, artificial intelligence (AI) allows computers to substitute their own. As AI research has progressed, it has been employed for a wide range of purposes, from defeating top chess and Go players to guiding self-driving vehicles. Artificial Intelligence encompasses a wide range of technologies, including big data, machine learning, robots, and deep learning.

"The capacity for computers to learn from experience, i.e. to adjust their processing in light of newly acquired knowledge," is one of the subsets of artificial intelligence known as "machine learning. Machine learning is defined here, although the field as a whole is not explained fully. Image identification, a frequent machine-learning problem, can be solved with the use of machine-learning techniques such as neural networks. Computers may be taught to recognize and group items using supervised learning. Because it understands that an octagon has eight sides, it will group all eight-sided items together. The system does not follow a predefined set of clusters or attributes of objects in unsupervised learning. When the algorithm notices that numerous objects have eight sides, it will construct clusters based on the commonality of those features. Learning by example is one of the most natural ways for humans to learn and is utilized in both supervised and unsupervised learning. A technique known as deep learning makes use of neural networks made up of multiple layers that are loosely modelled after the brain and that are capable of identifying and merging smaller, simpler patterns in order to recognize larger, more complicated ones. Sound, pictures, and other data can be analyzed using the technique. Predicting the outcome of court processes, precision medicine (medication tailored to an individual's genome), and transcribing language into English text with as little as 7% accuracy are just a few of the many applications for deep learning that exist today.

Robotics is the study of how robots work and how they might be improved by examining the relationship between what robots can sense and what they can do. It is a combination of computer science, mechanical engineering and electrical engineering, and is one of the key technologies utilized for automation in the modern world. Machine learning, computer vision, and natural language processing all fall under the umbrella of AI.

Advances in programming, sensors, AI, and robotic systems have greatly improved the intellect, senses, and dexterity of robots during the past decade. As a result, robots are now more adaptable, more compact, and more interconnected than ever before. As a result, it is now safer for humans and robots to operate together, and the number of applications for robots has grown dramatically. The advancements in technology have made it possible for robots to enter the service industry, which was previously considered unthinkable. Increasing technology advancements and lower prices are projected in the future for robots. As a result, robots is projected to see significant growth.

Robots are widely used in industrial and warehouse environments, such as for picking and placing, welding, packaging, and palletizing, because they have already surpassed human gross motor skills. Even some of Amazon's warehouses have been totally mechanized by robots. Technology, on the other hand, lags far behind in terms of fine motor abilities and dexterity. The human cognitive system is strongly intertwined with manual skills. There is still a lot of work to be done in the field of sensorimotor control of small and deformable objects. Miniaturized actuators and optical and tactile sensors, which perform at a level far below that of humans, limit the dexterity of robots. As a result, robots don't yet have the same degree of freedom as human hands, and present control systems aren't equipped to handle the variety and complexity of manual activities. There are, however, a number of anthropomorphic robot hands with human-like skills available. Navigation has already surpassed human capabilities thanks to advancements in machine vision and machine learning.

Precision locating and navigation are now possible with the help of cutting-edge GPS devices and massive amounts of spatial data. Google Maps and other navigation apps, for example, already employ these functionalities. In spite of breakthroughs in computer vision, robot mobility, and particularly autonomous mobility, remains poor. However, responding to new and dynamic situations remains a significant barrier for autonomous movement in static contexts (e.g., particularly designed warehouses).

A Shift in The Nature of Work

For the sake of this section, we will focus on the individual tasks that make up a job, rather than the complete job as a whole. This is due to the fact that jobs encompass a wide range of responsibilities, each of which has a unique relationship to the current technological capabilities. As a result, some tasks can be automated, while others cannot. The first step in assessing the impact of automation on jobs and labor is to determine which specific tasks can be replaced. Increasingly, technology has the ability to take over a wide range of tasks. A long period of time has passed since robots were capable of taking on some non-routine duties that had previously been outsourced to humans. Routine tasks have a high degree of substitutability, and that substitutability will only grow in the future. There is still a lot of room for human intervention in non-routine jobs, but it's confined to a few specific applications.

Many activities can't yet be replaced by machines, and machines can't execute a variety of tasks simultaneously. As a result, they are frequently incapable of filling in for full-time employees, who typically perform a variety of bundled tasks. It is better to focus on the specific functions of a job in order to establish the job's substitution potential.

Researchers have found that technology has the potential to replace a considerable portion of every work, regardless of industry or socioeconomic status.

In jobs that can be completely automated, the majority of the work is likely to involve repetitive manual and cognitive tasks that require little human interaction or manual skill. Typical examples include sewing-machine operators and employees at a retail store. The second factor to consider is that vocations with a high probability of being automated involve not only repetitive manual and cognitive work, but also human interaction or unpredictable/high-precision physical activity. Examples of highly automatable jobs include manufacturing and production because of their high level of manual routine work, and administrative support jobs because of their significant dependence on information gathering and processing (World Economic Forum, 2016). Transportation and material handling, as well as food and lodging services, are other occupations that need a lot of routine manual labor. Among all categories, this one has the most automation potential. Finally, the lesser the job's automation potential, the greater the share of non-routine tasks will be. Human engagement (requiring natural language processing and emotional and social capabilities), creative thinking, logic reasoning/problem solving as well as high-level dexterity or movement are all factors that boost this effect. A job that relies solely on these kinds of qualities is completely unaffected by automation. As an example, a choreographer's primary role is to produce choreography and engage in human connection with stakeholders and dancers in order to bring their work to life. A dentist, on the other hand, must have exceptional dexterity, sensory perception, emotional intelligence, and interpersonal skills in order to work with patients. As a result, automation is virtually nonexistent in either profession. There are still a lot of occupations that fall somewhere in the middle. Routine and non-routine tasks are included in this category. As a result, some of them can be automated in this way. When it comes to investment bankers, cognitive skills are critical, but a big portion of their work is acquiring and interpreting data, which can be automated. There are several legal professions where this is true. These kinds of professions are unlikely to go away; instead, technology will be used to enhance worker productivity and product quality.

3. Innovating New Roles for Workers Who Undergo The Digital Transformation Process

Instead of simply adapting to new technology, rethinking the entire ecosystem of Work is required if we are to truly reinvent it and fix the distortions that exist right now. It's no longer enough to simply do the same old thing. Currently, the workplace is more cognitively demanding, reliant on teamwork, more technologically reliant, more time constrained, and less geographically reliant. In today's environment, you may expect to work for a company that is likely to be substantially different from what it was in the past due to changes brought on by new technologies and competition. In today's world, organizations are becoming more agile, less hierarchical, more focused on identifying value from the customer's perspective, more aware of dynamic competitive requirements and strategy, and less likely to give lifelong jobs and job security.

Is It Feasible to Reimagine The Workplace?

A "Reinvention of Work" can only happen if three things are put in place:

- Changes in labor law that help redefine what work is, and reforms in the protection system that prevent abuses. Work contracts are divided into two broad categories under current law: those involving services and those involving labor. With a particular amount of labor agreed upon, one party enters into the loco conductio operarum contract. When one party promises to provide the other with the outcome of their labour in exchange for money, it is called a loco conductio operis contract. This division between employment and other kinds of service contracts is still present and may be traced in most legal systems (generally linked to freelancers and self-employed). It's not clear what this means. In exchange for a salary, I normally agree in an employment contract to provide a specific amount of Work over a specified period of time. For these contracts, the master/servant relationship and the labour movements of the 18th and 19th centuries have had an enormous impact. The struggle against the so-called piecework contracts, in which employees were paid based on the quantity of units produced, was one of the pillars of the social reforms requested. This bolstered the argument that workers should be compensated for their time, not their output. This sort of contract has historically been the subject of social conflict since it enjoys a higher level of legal protection than a service contract does in the majority of countries. Employees were previously thought to lack bargaining leverage, whereas self-employed professionals were regarded to have them. More and more people are finding work as a result of changes in manufacturing methods and the rise of the contemporary business. White collar jobs have taken their place in the workforce, moving away from the factories and into a new generation of jobs. It has been reluctant to adapt, but in general, legislation has acknowledged some amount of responsibility for the pursuit of objectives for a limited number of high-ranking executives. It is fairly uncommon for the legal definition of "executive" and the level of real delegation and management to be at odds. It has become increasingly difficult for blue-collar workers to have a significant impact on the total workforce as a result of automation and outsourcing. Companies are increasingly turning to alternative collaborations as a means of reducing labour costs as they seek to increase productivity. Some tasks have been delegated to private contractors. Individuals opted to be self-employed freelancers rather than be bound by an employment contract, so sacrificing some security in exchange for a steady stream of income. Pseudo self-employed workers have begun to emerge in response to this development, which means they lack the bargaining power of self-employed professionals but must accept non-employment contracts in order to survive. This pattern has been more pronounced in recent years, as evidenced by the rise of the "gig-worker." For this group of workers, legislation has been attempted in several nations, with various degrees of success. To put it another way: Instead of safeguarding lower-paid workers, front-line workers, and entry level positions, these contracts are now protecting the richer members of the workforce. A new labour law that recognizes the many shades of grey between "employment" and "freelance" activity, as well as the distinction between the two. It is possible to either recognize the same level of protection for all workers or construct a system that relates declining protection to the genuine bargaining strength of each employee.

- Changes in accounting rules have redefined the role of work in the performance of the business. One of the most important investments you can make. Companies argue that their most valuable assets are their employees, and you've heard it time and time again. Unfortunately, this isn't accurate from an accounting standpoint. Employees have no say in how their employers operate. They're not considered assets; therefore, they're just thrown out. Expenses for salaries are just a line item on the balance sheet. Organizations have changed in ways that accounting concepts have not kept pace with. And this raises a variety of concerns, as well. An accounting reason to invest in a robot is that it can be classed as an investment and depreciated over time, with the added benefit of eliminating a "fixed cost" right away, which makes the investment worthwhile. When it comes to accounting, there are no rewards for investing in people's professional growth and development. This problem is made worse by still another factor. In terms of corporate control, one can distinguish between fixed expenses (e.g., wages and benefits) and variable costs (e.g., consulting fees) (which instead is conceptually part of my variable costs). An accounting change that ensures the correct worth of "Human Capital" for a company is implemented. While this may seem like an easy task, the repercussions (such as changes to taxation) are numerous and complex. As a first step, financial analysts and venture capitalists should adopt a standard set of intangible metrics for evaluating investments in people. But to ensure that smaller organizations can also adopt these new practices, the GAAP must be changed.

- A shift in value perception necessitates the most drastic action, as it will necessitate a complete rethinking of the job market and the standards by which compensation are determined. Changes in labour laws and accounting rules are expected to benefit this large project. Those with bargaining power (typically assisted by executive salary data providers, often hired by CEOs to "pump" remuneration) and those without it (often relying on collective bargaining and minimum wage restrictions) are poles apart. One of the most significant effects is that demand and supply are not syncing well, which has a ripple effect that includes migrations (both within countries and beyond countries), challenges in education systems, etc. To truly grasp the concept of value, it is necessary to reimagine how we approach the design of work. Due to their "locked-in" focus on costs alone, relatively few organizations these days examine the "Potential Value" that roles and functions can perform within an organization. Companies and organizations all over the world stand to gain a great deal from this.

New Technology Can Be Put to Use in The Real World

Automation and streamlining of numerous processes will be made possible thanks to new digital technologies such as artificial intelligence (AI), machine learning, and robots, which all fall under the umbrella term "Practical exploitation of new technology." As a result, certain low-skilled, menial labor may be replaced, which means that people in this category may need to undertake reskill training. The usage of digital technology, such as artificial intelligence (AI), may have a significant impact on the way corporate boards operate.

The "gig economy" is made possible by the internet economy on a broader, more ordinary scale. Worker opportunities around the world will be made available to them thanks to the growth of "internet labor platforms". While this may be true when hiring workers, the cost of labor will frequently be a deciding factor (everything else being equal), which means that low-skilled individuals in particular will have to work at minimum pay in countries that use this method. While in countries with low labor-union density, where unions have the capacity to set minimum wage requirements for their members, salaries for workers risk falling into a "race to the bottom". These "gig economy" labour unions, however, are not clear whether they will be able to maintain their status, as this has proven difficult so far. Furthermore, in some countries, a large density of labour union members may not be achievable due to political or other reasons. Forcing an increase in the global standard of living has also been proposed, but it is still an extremely divisive issue, with many debating whether or not it is even possible to accomplish. Obviously, this is going to be a hotly discussed topic in the coming decade. The diffusion of information in the academic community will likewise become more thorough and openly available as a result of the digital revolution. More research into the benefits and constraints of the gig economy will be stimulated, as well as a greater understanding of this complex problem by the general public.

The Digital Welfare State's Function

The digital transformation will have a varying impact on welfare state functions in different countries (depending on how advanced/mature welfare is in each country), and all welfare state sectors will be affected. Companies will be encouraged to automate as much of their work as feasible or purchase services on gig markets in order to pay as little as possible in wages (and by extension, less tax). Workers may be forced to choose between jobs that can't be fully automated and jobs that require new skills learned in an increasingly technologically advanced society, which could lead to a more divisive labour market. It does not alter the fact that the "gig economy" is gaining traction in the workforce. Because of the "gig economy," wages may be driven down in some countries to the point of becoming uncompetitive.

Because of this, the state's capacity to collect substantial tax revenue from labour will be harmed, and this will only get worse as the number of "gig patients" increases.

Consequently, it will become increasingly difficult to maintain the "social contract" as the ability to give the same level of welfare as now is expected to decline significantly.

Digital Upheaval of The Established Order

Additional broadly, the "gig economy" and the "planetary labor market" have less to do with creating more work opportunities across geographic borders, but more to do with enabling firms to operate across countries at a little cost through the use of digital technology. Disillusioned low-skilled workers who are unable to reskill themselves and/or are unable to take on "gig employment" may benefit from a new "point-based" economy in the digital era. Through the "gamification" of their pastimes, this new economic model would enable this demographic to receive some sort of "trade-in" benefit. This will ensure that this group of people is not kept completely passive, while simultaneously allowing firms to profit from the "work" done by this group of people. Such an economic shift could lead to the rise of "alternative" and "virtual" currencies that are not too different from cryptocurrencies. Privacy concerns will always exist, and any regulatory measures taken to address them would undoubtedly have a significant impact on the job market. New legal frameworks, like GDPR, suggest that certain sorts of vocations, such as digital strategy consulting and analytics and software architecture professionals, will be in higher demand in the future. Increasingly, consultants will need to work more closely with data scientists as analytics technologies become more important.

Where To Go From here?

The abilities of different sorts of specialists are not necessarily interchangeable with one another. It's possible, but the talent held is often unique, and different competences may be more suited for certain activities than the ones listed here. Data scientists, for example, who are looking for "gig employment," could be forced to promote themselves more aggressively in order to gain recognition for their specific specialty and stay one step ahead of the competition. Instead, consultancies might grow their operations to include data scientists in their team of professionals to a greater level than is done presently, as well. Individual groups of society are disproportionately affected by digitalization's impact on inequality. Despite Castronova's concentration on young male low-skilled employees, other groups may suffer as a result of this policy as well. It is possible that the "digital gender divide" will widen due to the increasing automation of traditionally female-dominated professions and the difficulty of retraining as a result of the underrepresentation of female students in STEM fields, which results in fewer females working in ICT-related fields. Therefore, societal attitudinal change is critical to ensuring that women have a future position and, if necessary, the same conditions as men to reskill themselves. There are many benefits to adopting a circular economy, including the development of new jobs and a more sustainable economy, but digital transformation also provides the preconditions. Specialists in component remanufacturing and product refurbishing, as well as product re-sellers and reverse logistics specialists, may benefit from a circular economy. Even more significant, though, is the potential impact that a shift to a circular economy may have on current corporate practices. Some examples include organizations moving from traditional workspaces to open-plan settings in recent years. In large part, this can be due to the rise of "knowledge work" (i.e., workers whose primary expertise is knowledge, e.g., scientists, programmers) in the modern labour market and the development of mobile technology. People are no longer as firmly tied to a single workplace as they once were, which has opened the door to new, less expensive and more space-saving options like "hot-desking," "open workstations," "group workstations," and so on. They may save money, but they could also have a negative impact on productivity and morale, especially when it comes to "hot-desking." This is because of things like noise levels and stress from conflicts and other sources.

It is possible that the workplace will undergo a radical transformation in a circular economy. The co-working spaces may become more like "Airbnb-style" workplaces, where enterprises can book an apartment or home for a few hours whenever an "office" is needed. With the advent of "knowledge work" and mobile technology, employees may be able to work remotely or in a co-working environment while in different places.

4. Change Is Inevitable in The Age of Digital Transformation, But It May Also Be an Opportunity to Stretch One's Boundaries

It is possible to increase a company's efficiency and profitability by implementing digital transformation correctly. An increasing use of technology can speed up business operations and help businesses better respond to market shifts. As a result, only a small percentage of organizations reap the benefits of digital transformation. One of the top three strategic priorities for any significant organization today is digital transformation, and many have already started the process without fully grasping what it means to be successful. Many times, digital transformation is a matter of survival because of the simple truth that technology has forever altered our lives and continues to do so. Change that resulted in the emergence of a new breed of companies with technology DNA capable of creating a better substitute for the numerous services and goods that the "old" world companies provided. Furthermore, these "new" businesses are quick to adapt to the shifting demands of their customers. Old organizations structured upon human processes are exposed by the new world's agility as being woefully unprepared to compete on an equal footing in the new game. Large sums of money are utilized to set up a theatre play for posing as a new world firm, but the clock to their demise is unimpressed by the show and continues to tick.

What Is Your Company's Greatest Opportunity and Worst Threat?

The digital transformation of firms, industries, and society is advancing at an ever-increasing pace. Digital transformation refers to the integration, use, and exploitation of digital technologies to trigger major changes in value creation, appropriation, and delivery. In the past decade, there have been several calls that becoming more digital is imperative to stay future-proof. Finally, the COVID-19 pandemic and the resulting closures of businesses all around the world showed the need for digital infrastructure as online shops and digital businesses navigated comparatively smoothly through the crisis. In contrast, firms that just started on their digital venture had to adopt digital means of communicating and creating value overnight and often struggled in comparison to their more digital competitors. While another pandemic is hopefully decades away, the need to become digital is not going away anytime soon. Hence, digital transformation will remain a key point on the agenda for executives to guarantee that their businesses stay future-proof. Integrating digital technologies to create an online offering was used by many small to medium enterprises during the worldwide pandemic. From one day to the other, many small firms such as restaurants were cut off from any income stream and managed to switch towards online ordering systems to deliver food to their customers. While this new means of value creation often were not enough to replace the entire income stream, digital technology helped these firms to at least create some income despite being unable to open their doors. Aside from forced digital transformation, there are also more deliberate (digitalization) journeys about local firms.

However, if unsuccessful, digital transformation can lead to a company's demise. Hence, executives nowadays are in jeopardy between transforming to become future-proof on the one hand and risking demise on the other hand. On the bright side, digital transformation is possible and with this book chapter, we try to equip businesses with the latest knowledge from academia, packaged in a way that is easy to grasp and helpful for business.

Connect With Organizations & Societies

While it may be your primary responsibility to take care of your family, you can't do it all on your own. This is especially true if you're providing care from a great distance (greater than an hour's travel). Health care providers and family members as well as friends will be needed to aid you through this process. Your ability to provide care will be harmed if you don't receive the help you require.

Prior to requesting assistance, you must have a complete picture of what your loved one is going through. Make a detailed list of all of the caregiving responsibilities you will be responsible for. As a result, figure out which things you can participate in (be realistic about your capabilities and the time you have available). The remaining items on your to-do list will necessitate the assistance of others.

Transforming From a Paper-Based To An Electronic-Based Society

Besides digital transformation, the notions of digitization and digitalization are also relevant and often overlap. The initial phase of digital transition, digitization, refers to the conversion of analogue data into digital data. Digitization shows the fundamental need for digitalization and transformation to some extent. In the real world, digitization can mean that paper forms, such as those used by employees to register their vacation days in a system, are replaced by digital forms. In the second place, digitalization refers to changes in processes that are being digitalized. Digitization refers to the second phase of digital transition and can be used to speed the ordering of goods, including online shopping as a new channel or to streamline internal process management. In many cases, digitalization serves as a bridge to full digital transformation. Digital transformation is the third and most pervasive phase of digital change.

Change in the way a company utilizes digital technologies to establish new means of value creation and appropriation, as well as a new digital business model, is a sort of digital transformation. The three criteria together reveal a clear route for firms to follow. Each of the three stages of digital change offers a unique set of possibilities and challenges. While cutting costs is the primary goal in the first phase, new ways of working with digital data open up as a result. There are a slew of enhancements and possibly new sources of value already possible in the second phase, as well. These two digital phases provide firms with the knowledge they need to move forward with full-scale digital transformation activities. A caveat is in order: these phases can't be treated as discrete stages, as pieces can overlap and stages can be skipped entirely (though this is unusual). The importance of technology in digital transformation is critical, but it is not the only factor in the process. If a company acquires new technologies but does not know how to use them in new ways to produce value, this is not a digital transformation and is likely to lead to bankruptcy. To acquire an advantage in today's global economy, it is no longer necessary to own and control distinctive resources like machinery, as was the case in the old economy (before 1990).

Most resources in the digital economy can be obtained for little or no cost. All of the most advanced artificial intelligence and machine learning algorithms are open source. Several open-source Python libraries, such as PyTorch and TensorFlow, are used heavily in Tesla's self-driving car mode of operation. To be clear, developing and training an AI still takes a significant amount of money and effort, but the "technical" elements that go into creating new recipes (i.e., innovations) are widely available. Everyone has the capacity to reproduce the same technology if they have the proper human capital (e.g., engineers and data scientists) and access to sufficient time and training data. In this situation, the importance of physical resources (in this case digital technology) diminishes, while the relevance of human capital that manages such technology rises. As a result, digital transformation is not about the technology itself but rather the ability to use technology to create value.

Opportunity To Invest in The Stock Market

Digital transformation is a way of doing business, not a project. Better business outcomes are ensured when agility and responsiveness are consistently maintained. Talent and cultural concerns are the main roadblocks to digital revolutions. Finance transformation, in my opinion, is the same as it always has been. It's time to get out of our comfort zones in 2020, since by 2025, the vast majority of financial transactions will be automated, whether we like it or not. This is the year to start thinking digitally first and foremost. A corporation must often begin a process of change from the top down. We are seeing more and more need for financial professionals to contribute value and provide insights because of the success of digital transformation. Financial business partnerships are in jeopardy if they don't add value, and they're nearly difficult to develop without digital transformation.

The finance department's digital strategy must be in line with the company's overarching digital framework and overall business objectives if it is to be effective. It's clear that finance requires a set of tools to help it perform better and better serve the business, which generally includes a combination of tools.

IT services that are delivered over the internet using cloud computing, such as applications for planning and procurement. An estimated 25% of financial apps are hosted on the cloud, according to recent research.

Automated software's programs are used to execute repetitive operations and automate processes such as P2P, O2C, etc. 30 percent of organizations utilize process robotics, and up to 80 percent of FTE work can be automated, according to a study.

To better serve customers, corporate intelligence and analytic systems benefit from the innovative use of visuals and interactive technologies known as visualization. More than 40% of firms are using or testing visualization tools, according to a recent study.

Advances in analytics are helping businesses answer numerous issues with deeper insights and more accurate predictions, notably in the areas of planning and profitability analysis. CFOs in the healthcare sector said they've made or will make investments in this area, according to a new study. Cognitive computing includes artificial intelligence (AI), machine learning (ML), natural language processing (NLP), and speech recognition (SRR). More than half of organization's are either using or evaluating new technology in this area, according to a recent study. Only 10% of CFOs polled said they were using it.

Digital ledger where transactions are validated and stored without a central authority is known as the block-chain. Only 4% of CFOs polled reported using this technology, indicating a lack of familiarity with it.

Models of Digital Businesses

It's the method a company generates and delivers value to customers, then converts payments into profit that is known as a business model. As a result, a business model is a description of how a corporation generates its revenue. When it comes to digital transformation and future-proofing, the business model is the most critical consideration. A company will eventually fail if it doesn't have a robust business strategy in place. Digital business models differ significantly from traditional ones in numerous important ways. Without the correct technology, many of today's most successful business concepts might not have been viable at all. As an example, consider a paper catalogue that tries to compete with Amazon's tens of thousands of products. This catalogue would have been too cumbersome for any customer to peruse in a reasonable amount of time. With digital technology (such as search and virtual product spaces), new opportunities to create value can be realized. Digital business concepts come in a wide variety of shapes and sizes. In the digital economy, even small and medium-sized businesses (SMEs) need to think about how they might compete with larger organizations (e.g., platforms). In addition to the fact that digital business models are more profitable, they are also essential for small businesses to compete with larger corporations and to remain relevant in the future.

Using the freemium business model, a service or product can be offered for free, with the opportunity to upgrade to a more advanced version for a fee. A free product or service is offered by the corporation, but customers can upgrade to a paid version for an additional fee. Customers may be lured to the website/platform/service with a free offer, but the company rarely makes money from them. Thus, the organization can begin making money as soon as users purchase new features. To be profitable, a video game company must offer a product that is superior to its competitors' premium offerings. Accordingly, the marginal (distribution) costs of serving a new consumer are nearly negligible in this digital business model. Adding new clients is a cinch once the offering is in place. In order to be successful, this business model necessitates a large target market in order to be viable as an independent entity (e.g. in the video gaming industry a specific niche such as middle age strategy games). The fermium business model is used by Dropbox. Although the basic service is free, clients can pay a membership fee to gain access to more storage space, as well as better sharing and access capabilities in the cloud. In order to use the free service, the user is limited to a small amount of storage space, which prevents them from uploading large amounts of data. As a result, Dropbox offers a variety of storage options, each with a corresponding increase in storage capacity. One of the many marketing automation companies that offers a free service but offers more advanced services for a fee.

Some people believe that nothing worthwhile comes for free, and this is especially true of digital technology, where users will have to pay for some services. Using this business model, a company provides a service or product for no cost to the customer. As a result, there are no "hidden" fees because the service is free to the client, unlike the freemium model, in which new features are charged for (see above). On the other hand, this model relies on at least one of the following: To begin, client data is used to generate value. Ads are also a source of revenue for the company. It is also possible to make money from commissions (e.g., affiliate marketing). Because of the wealth of available data, these models can be integrated to produce even more precise results for customers. These business models are particularly scalable since the marginal cost of gaining new consumers is nearly $0, making it possible for them to grow rapidly. Google's search engine and social media platforms like Facebook, Instagram, and TikTok are all good instances of free products and services. All of them are free, but they make money by selling customers' information to advertisers. On a smaller scale, free web blogs on specific niche areas might generate substantial revenue if enough people discover the website.

Rather than delivering actual goods, firms today allow customers to use their items by paying for access to them. As a result, with this business model, corporations no longer sell the product itself but rather provide customers with access to it via a subscription or per-use cost. The business concept for renting and leasing can be seen in this model. Contract lengths are typically longer when renting or leasing an item than when using a subscription or pay-per-use business model (compare weeks against months or years). Providing a product or service generates revenue for the business. Profitability comes from establishing a steady customer base that makes repeat purchases rather than having to worry about fluctuating product sales. There must be a large consumer base and enough products to keep up with demand at peak times. In addition, making the switch to a subscription model takes time and money. When Netflix charges a monthly subscription price, it allows users to access the complete film library at any time. In addition, carsharing services charge a fee for the actual usage of the vehicles they provide. In addition, automotive firms are also offering subscription models in addition to leasing and selling vehicles. If you're looking to expand your current product line, this business strategy could be a good fit for you.

Is It Worthwhile?

Finally, a critical consideration for companies considering digital transformation is whether or not it is merely another management trend that will go away. It's difficult to say for certain whether or not digital transformation is a long-term trend, but the overwhelming evidence points to the contrary. Digital transformation has the potential to be viewed as a second industrial revolution, with far-reaching consequences for people's daily lives. Preparing for the digital age is as important as changing to steam, electricity, or communications in the past for becoming future-proof. In addition, many organizations are currently undergoing digital transformation, and CEOs are expressing their efforts as a must to be future-proof. Digital transformation is therefore worthwhile only to level the playing field in this sense. Digital business models are at the heart of a number of the world's most successful organizations today. These digital business models have been embraced by the majority of Fortune 500 companies in the United States and China, including heavyweights such as Facebook and Amazon as well as smaller players like Apple, Netflix, and Google. However, small and medium-sized businesses (SMEs) are also changing their business strategies. Small and medium-sized businesses (SMEs) can benefit from digital transformation by finding new methods to generate value and interact with their consumers.

If you want to be ready for the next digital decade, you need to embrace digital transformation. It's not just a fad. Digital business models, on the other hand, have a huge profit potential even for smaller companies and allow for whole new ways of creating value. Last but not least, companies should not be frightened to take on the challenge of digital transformation because it is good to their bottom line. Although some businesses will fail in their digital transformation efforts, they may be worth it in the long run if done correctly.

5. Digital Transformation Is a Friend Or A Threat?

We are moving ever closer to a world where artificial intelligence and automation are commonplace at work. Economic development should be fueled by increased productivity as a result of digital transformation. It's easy to imagine a future where machines forget who they work for, as depicted in the film Terminator. Humans in the workplace may suffer the most in economies with the greatest potential for growth. As a result, ensuring that the workforce of the future is equipped with the necessary digital skills requires a considered and sensitive approach. Today's defining challenge and opportunity is the ability to develop a human-friendly innovation economy.

Identifying the root causes of this problem is the first step in dealing with it. The challenge is that many of the inputs in an innovation economy have yet to be developed. However, an examination of the past and present indicates that technology has had two effects on our work: to enhance and replace. For a human-friendly innovation economy, both dynamics must be taken into account Technology has always superseded human labour, starting with the latter. A human desire to go beyond one's limits can be traced back. The engine, like the wheel, allowed a single person to push something that previously required tens of people to raise. Digital progress has exploded in the last few decades in a way that is both exciting and alarming. A large part of this is due to a lack of education, since we often have a fear of the unknown. But there is also a very real and large human labour cost associated with this issue. For example, manufacturing jobs in the United States fell from 16 percent in 1996 to 8 percent in 2016. This is a dramatic change. Even though many people are impacted by anxiety, we shouldn't let it take over our lives. Progress has its advantages, but we must first prepare ourselves to reap those benefits. The second historical impact of technology has been to increase human productivity. The advantages of technological progress can't be overstated, even if the dread of widespread unemployment sometimes overshadows them. In the digital age, businesses are able to do more of what their customers want – faster, better, and cheaper – as a result of digitization. Employees can become "superheroes" thanks to digital augmentation: a line worker with robot support can manufacture more items; a lawyer with real-time case analysis can generate a stronger argument; and a biochemist with data-informed genetic models can heal more diseases. In our current and future digital world, these employees are flying through production hurdles.

The Shadowy Side of Things

Digital transformation is one of the most publicized marketing projects of the past decade. As a process rather than a task, it has evaded criticism since it is considered as the required antidote to a disrupted and computerized world. Instead, then being transformed, the majority of businesses are "transforming." It's easier to put the blame on poor implementation than a faulty strategy if things go wrong in this direction. However, many shops and brands see digital transformation as more of a traffic grab than anything more. Customer experience may be improved through personalization and marketing automation because their key performance indicators are focused on driving traffic and converting it into customers.

The unintended consequence of digital change is a weakening of brands. In most cases, brands and the corporations that support them are not technology-driven. Outside partners are necessary for them to establish their own internet properties, as they lack the resources and experience to do so. They almost never create anything that naturally reflects or develops from their brands.

Is there another side to the story? Disruptive technology platforms have spurred many companies to adopt digital transformation. It goes without saying that it is tough to compete with a company that also provides the backbone of your operations successfully.

Every online retailer is required to provide free; no-hassle returns. Due of people's tendency to make poor online purchases, convenience isn't the only factor. This has never been an issue for brick-and-mortar stores because they have salespeople on-hand to assist customers. It is likely that enabling contacts with customers will lead to increased sales, but it will also assist minimize turnover significantly.

Brands can't reap the benefits of digital transformation if they are all given the same value. Brands die because of commoditization, which eliminates differentiation. Consumers want a better and more educated purchasing experience, which means marketers must find new methods to use technology to differentiate their companies.

Risks to Security

Data, systems, and people are being attacked by cybercriminals who are taking advantage of the digital transition. The volume of data that must be recorded, saved, managed, and safeguarded grows as more and more transactions and interactions are conducted via digital channels. Customers' behavior, cross-sell and up-sell efforts, and demand forecasts can all be improved with data. Cybercriminals are also interested in the same information. In addition to facilitating identity theft, it provides useful information for phishing efforts, fuels the development of business email compromise attacks (e.g., to divert payroll funds or invoice payments), and much more There are a plethora of data breach statistics accessible, and a thorough list would necessitate a full blog entry. In the first three months of 2020, 8.4 billion unique documents were exposed, with the majority coming from just 11 breaches of at least 100 million records each. There were more breaches in the first three months of 2020 than in the first three months of any year from 2013 to 2019. The situation is deteriorating. New systems and apps to support digital transformation activities and older systems modified to connect to the internet world, while businesses rejoice in their increased asset base to deliver digitally, cybercriminals rejoice in the ever-growing attack base accessible for compromise. An increasing number of ransomware attacks are becoming more and more destructive, with critical systems hacked at the most inconvenient times, data stolen and threatened for resale to increase the likelihood of paying a ransom, and other types of extortion designed into attacks to increase the likelihood that the cybercriminals will profit from their misdeeds.

Personal and professional lives are intricately connected in social media platforms that cater to a global audience, as seen by the vast online professional networking services (such as LinkedIn) (e.g., Twitter, Facebook, Instagram). It's hardly surprising that hackers are active on these services, given that they have access to billions of individuals via computers and mobile devices. A perfect storm has formed for cybercriminals, with everything from fake friend requests to scraped profiles impersonating someone you know to the use of fake profiles to build rapport and pretext an attack to fake job offers.

Infrastructures Utilizing a Combination Of Public And Private Cloud Services

More and more companies are moving to multi-cloud IT architectures, often supported by more than one cloud service provider. SaaS, PaaS, and IaaS are all examples of this type of service. Cloud computing, regardless of the sort of cloud being used, presents a significant risk, especially when many locations, services, or vendors are involved. It's not just that cloud storage might be a security risk, but that poor cloud management methods can also lead to overrun costs and data privacy issues. Governance of cloud settings is the most common danger we encounter here. Which cloud service provider should I use? Is there a certain methodology that you'd like to Thresholds for [development] environment formation, utilisation, size, etc., in order to optimise use. Rather of dealing with governance difficulties after deployment, it is considerably easier to address them at the outset. Since IT services have traditionally been purchased from company-owned and operated data centers, IT has been responsible for overseeing the procurement process. There are no longer any architectural or security reviews required for cloud services such as PaaS. It is important for IT and business management to control which services are turned on and available for use.

As a rule of thumb, cloud services should undergo proper architecture and security examinations before being certified for use in the company, Smith argues. Guidance and guardrails must be developed before any public cloud vendor solutions can be offered to the enterprise, including continual monitoring of all usage. The efforts of business users to acquire and utilize new cloud services necessitate a joint effort between IT, cybersecurity, and legal.

Supply Chains and Channels for Sales In The Digital Age

End-to-end digital connection, cloud services, blockchain, robotics, autonomous vehicles, and advanced analytics tools are among the technologies increasingly relied upon by businesses to improve and manage their supply chains. Digital transformation of the supply chain may improve efficiency and visibility, decrease errors and costs, strengthen communication with business partners, and optimize operations. Also, there are concerns, such as data loss, that can arise. The parties involved in B2B digital services might apply a variety of risk mitigation measures. This includes the development of thorough business agreements with partners that deal with the numerous risks and duties involved. Security and privacy measures for the transfer and storage of data can also be implemented by companies. These B2B connections are frequently monitored by businesses in order to guarantee that regulations and procedures are being followed. All participants in the digital supply chain should be held to industry security and privacy rules and standards, as recommended by best practices.

Digital sales channels including ecommerce, email, SMS, mobile apps and online events are also becoming increasingly important for companies to reach consumers and prospects.

There is a risk of not having a clear multichannel strategy, or, if going entirely to digital, of not having a strategy for allowing partners, customers, and other end users to make the move. In the absence of a well-defined strategy, businesses may find themselves in a continually shifting priority position where effectively none of the channel's proceeds. When it comes to building various digital channels, infighting can occur. An essential mitigation method is to have a single leadership team responsible for and accountable for all of the various channels.

The Interconnectedness of Everything (IoT)

In order to track the location of assets and keep track of equipment performance, manufacturers, healthcare providers, retailers, and others have begun adopting IoT devices in large numbers. Benefits include improved supply chains and factories, better equipment and product maintenance, better customer experiences, and lower costs due to the avoidance of lost goods. These are just a few of the advantages that could be realized. The dangers, on the other hand, are substantial. IoT methods present multiple entry points for hacking, including linked devices themselves, which have already been accused for distributed denial-of-service assaults. Everything from HVAC systems to servers and other IT equipment to automobiles, lighting systems to thermostats and other home appliances can be included in a networked environment. A networked device's vulnerability can be reduced by separating it from other networked devices, or by putting it in a separate network entirely. Other best practices include mandating IoT device manufacturers to provide means to keep devices up to date and secure; and monitoring corporate networks to detect IoT devices for suspicious activities.

Automated And Data-Driven Decision Making

Automating labor-intensive and time-consuming manual procedures is a top priority for businesses looking to increase productivity while lowering mistake rates and overall operating expenses. Artificial Intelligence (AI) and robotic process automation (RPA) have the potential to revolutionize commercial operations, but they can also present hazards. With analytics, AI, and machine learning (ML), one of the most important risk factors is the datasets used by data scientists to train their algorithms.

When it comes to dealing with big data collaborations, Smith recommends drafting well-crafted contracts and restricting data sets to the least amount of information essential. Smith also suggests using anonymized data whenever possible. Some of the dangers of automation can stem from a lack of ability to grow or meet goals quickly enough. A lot of things are changing in the automation industry right now. From process outsourcing through Lean and Six Sigma to RPA, it's been a long journey. RPA and AI are now working together to tackle difficult business issues. New possibilities and use cases are being opened up by this congruence of AI and RPA, such as intelligent document processing with 175 billion machine learning parameters or the use of neural networks and deep learning to detect anomalies in transactions. Ehsan advises organizations to begin preparing for automation as early as possible and to involve both business and IT stakeholders in raising awareness of the potential advantages, capabilities, and applications of automation. When this is done, they should begin with tiny, quick, and short-term pilots that emphasize the advantages.

Customer Satisfaction Is Improved

There has never been a time when customers had such high expectations. Customers, on the other hand, are looking for services that are tailored to their specific demands. When it comes to duties such as accessing, upgrading, or deactivating and reactivating software, they don't want to deal with dealing with humans. One of the most obvious benefits of digitalization in business is the ability to apply sophisticated digital tools and strategies, such as location- or industry-specific license modifications and automated processes. Customers are more satisfied as a result of these computerized procedures. The drive to better serve customers is at the heart of any digital transformation. When it comes to capturing customers at the top of the funnel, this is true whether it's addressing search inquiries or improving the user experience of your website. In SAP's digital transformation executive research, nearly all business executives say they have matured digital transformation initiatives in place to improve the customer experience. More than two-thirds of those polled say that digital transformations aimed at increasing customer happiness and loyalty have been highly successful.

Considering that two-thirds of a company's competitive advantage derives from the customer experience, organizations who fail to priorities digitizing the customer journey will soon be left behind.

Agility Is Boosted

When personnel are well-versed in digital strategies and technology, businesses are better equipped to respond quickly to market shifts. More than two-thirds of firms, according to CIOinsight.com, place agility at the top of their list of priorities when tackling digital transformation. Businesses must be ready to respond quickly to changes in the digital landscape. Customers will continue to want faster service as technology progresses, which will lead to innovation races and more rivalry. We must be willing to experiment with new technology if we are to remain competitive in a rapidly changing world. This will allow us to adapt more quickly, which will allow us to prosper in the face of shifting public expectations. Since the digital landscape has evolved so quickly, the public sector has access to a wide range of software options. There are a variety of technologies that can help businesses innovate and adapt, such as virtual reality, artificial intelligence, blockchain, and natural language processing. Keep in mind that merely implementing new technology does not inevitably result in increased innovation and agility. For new ideas to be tested, you need to foster a culture of innovation and cooperation.

Benefits for Industries

New approaches, skills, and sources of income can be created by combining practices and ways of working in the digital transformation business. Because of the time reductions in the procedures, it lowers expenses. It facilitates mobility and remote communication, which decentralizes production. It boosts productivity and efficiency in the workplace. new business options and revenue streams are opened up, allowing for the development of new products and services. It improves the company's ability to respond quickly to changes in market demand. A competitive advantage is created for the company since it can improve the quality of its products. It promotes an innovative work environment and helps the organization prepare for potential disruptions. It facilitates departmental communication, which enhances integration and collaboration within the organization. by improving data analysis, it empowers decision-making (Big Data), As a result, the awareness of systems and the curiosity of specialist specialists are sparked by the fresh talent that is brought in.

Enhanced Data Accessibility

We may gain greater visibility and insight into our data through an efficient digital transformation plan. We can get a clearer picture of our business's performance by modernizing our data collection and handling. In addition, this data is made more widely available, so that anybody can have access to a project's data. Ideas and information can be disseminated more quickly and efficiently when everyone has equal access to it. We can better understand our customers and the people who use our products and services if we have better access to their data. So, as a result, we should expect to see a higher return on our coordinated efforts.

Improved Workplace Culture

In the meantime, let's mention that digital transformation helps build a strong corporate culture. It's a give-and-take relationship, but the building of a digital workplace is essential to digital transformation. Organizational cultures that are more flexible and responsive to the demands of the digital world tend to have higher levels of productivity and employee involvement. This is especially critical in the public sector, which serves the greatest number of people and provides the most essential services. In order for a business to remain relevant in the digital age, it is important to have a strong organizational culture in place. By creating an environment that encourages and promotes new ideas, the company encourages its employees to continuously enhance their skill set. Cultural change is closely linked to digital development. They're actually two sides of the same coin that are mutually dependent on one another in order to thrive and persist. An overview of how to implement cultural change in your company is necessary if you want to learn further.

Foreshadowing Of What Is to Come

Businesses are being reimagined by digitalization and gaining novel competitive advantages, yet firms are hesitant to alter their established business models. Many top executives' catchphrase in the last few years has been "digital transformation." Fewer than half of the organizations claiming to be digitally transformed have integrated digital activities into their main strategy to date. We live in a digital, global, and hyper-connected society, where new players in the market, off-site mobility, and constant communication have a significant impact on our lives. To remain competitive in today's technologically advanced world, organizations must undergo digital transformation. This is no longer a choice.

6. Transfer of Data

In any digital transition, one of the most challenging and time-consuming parts is moving data. In order to have a successful digital transformation, most firms overlook the importance of data migration. A lot of companies aren't aware of the difficulties or stages necessary. The process of migrating data is iterative. Some people feel that data can be moved using "lift and shift" (auto-magical) methods, while others believe that the planning and heavy lifting will be handled by someone else on the team.

Be prepared for a long and difficult technical process that necessitates numerous tests (think multiple rounds). When a large-scale migration is handled by a project team that lacks the necessary resources, the outcomes might be substandard and have a substantial influence on the larger undertaking. It is only when real data is used to test the new system configurations that the shortcomings of the system may be discovered. Introducing data too late in the project cycle prevents early detection of system flaws when it is not a well-planned component of the project. Project delays and costly design/build/test rework cycles can and will be caused by late modifications to the new system.

It's also possible that "dirty data" could have an effect on the outcome. Trash in, garbage out happens if data preparedness is not adequately planned and performed. IT projects on a wide scale promise a variety of benefits, such as business intelligence, data analytics, and the Internet of Things, for example. A lack of quality, consistent, or consistent data will render these technological tools ineffective.

Most individuals agree that the quality of your data-based insights and analysis is directly related to the quality of the data you are employing. Garbage in, garbage out, as the saying goes. If you want to cultivate a culture of data-driven decision-making in your organisation, data cleaning, also known as data cleansing and data scrubbing, is an essential step.

Cleansing The Data

A dataset's data can be cleaned up by repairing or deleting any errors, corruptions, malformations, duplicates, or incomplete records. It's easy for data to be mislabeled or duplicated when you combine data from several sources. Even if the data appears to be correct, the results and methods are unreliable. There is no one-size-fits-all approach to the data cleansing process because each dataset has its own unique set of requirements. Establishing a template for your data cleaning process is critical, as you'll want to ensure that you're following it exactly each time. The process of removing data from your dataset that does not belong there is known as data cleansing. Converting data from one format to another is known as data transformation. A data wrangler, or data munger, is a person who transforms and maps data from one "raw" data type into another for storage and analysis. Listed here are the fundamentals: -

- Remove any unnecessary data from your dataset, such as duplicates or irrelevant data. During data gathering, there is a high probability that there will be a lot of duplication. Duplicate data can be created by combining data from different sources, scraping data, or receiving data from clients or other departments.
- When you make an observation that has no bearing on the problem you're trying to solve, you've made an irrelevant remark. It's possible to eliminate observations from your dataset that don't pertain to millennial customers in order to conduct an analysis on that data. As a result, your core goal can be better served and your dataset will be more manageable and more performant.
- Structured errors are found when performing measurements or transfers and discovering unusual naming conventions or typographical problems. An incorrectly designated category or class can result

from these inconsistencies. Not applicable and N/A are both examples of the same category, and should be treated as such.

- When evaluating data, you may come across one-off observations that don't appear to fit at first glance. You can improve the performance of the data you are working with by removing an outlier if you have a good cause to do so, such as incorrect data entry. However, there are situations when the presence of an outlier proves a notion. The mere existence of an outlier does not imply that it is erroneous. This is a necessary step in order to verify the accuracy of that particular number. It is possible to remove an outlier if it is irrelevant or a mistake.

- Missing data cannot be ignored since many algorithms reject it. Missing data can be handled in a few different ways. Although neither option is ideal, it is possible to explore them both. There are a number of ways to deal with missing values in your data, but you should be aware of this before you remove them. The second alternative is to fill in the missing numbers based on other observations, but there is a risk of losing the integrity of the data because you can be operating from assumptions and not actual observations. To get around null values, you can also change how the data is used.

Poor business strategy and poor decision-making can be informed by erroneous or "dirty" data that leads to false conclusions. The moment you learn your data doesn't hold up under scrutiny after drawing incorrect conclusions in a reporting meeting can be embarrassing. In order to get there, you must first cultivate an organization-wide culture of data quality. If you want to develop a culture of data quality in your organisation, you need to record the tools you'll use and the meaning you place on data quality.

Stations For Storing Personal Data

To improve control efficiency and minimize expenses the enterprises and organizations strive to optimize existing business processes. The most important asset is information availability and relevance of which are the key factors of success in competitive environment. This situation stimulates large-scale development of data storage technologies. The state-of-the-art data storage system ensures reliable storage of information resources and access to them. The system includes disc arrays, access infrastructure and data storage control software systems here.

Preparation of Data

When it comes to making sure that a company's artificial intelligence (AI) platforms are fed with accurate and relevant data, this is what we mean when we talk about being data ready. The tech industry has responded to the promise of AI/machine learning (ML) by increasing its attention on the development of toolkits that enable the creation of faster, better, and more automated models. That's where the majority of those efforts have concentrated. Consequently, data is not properly understood, evaluated, and verified before it is employed in the construction of ML models. Businesses must embrace data readiness to close the skills gap between data scientists and business analysts, which often accounts for 60-80% of their time. The quality and relevance of a model can only be determined by the data it is trained on. To solve this difficulty and ensure that context-driven high quality is employed in constructing ML models while speeding the ML lifecycle, data readiness is an essential part of the process.

AI is increasingly being used by a wide range of companies, from merchants to automakers, to develop new services and products and provide more personalized service. An increasing number of companies are embracing artificial intelligence (AI) as a way to deliver more personalized customer experiences. However, AI is only as good as the data it relies on. As the amount of unstructured data grows, the work of preparing it becomes more difficult. From social media posts to chat chats, unstructured data provides businesses with real-time information that may be used for decision-making. Formal research just cannot answer the type of raw information into a company's operations that unstructured data provides. It's also widespread: up to 80% of company data is unstructured, according to IBM. However, unstructured data can be a problem. It doesn't fit neatly into any predefined category. AI may be trained through machine learning if firms invest more in improved methods of acquiring and preparing that data.

Customers expect businesses to use first-party data more wisely in order to provide a more tailored experience. Third-party cookies are essential for businesses to create personalized experiences across the web, and technology giants such as Google have taken steps toward eliminating them. The Google Chrome browser will no longer support third-party cookies as of 2020, for example. Third-party tracking is taking another hit from Apple's new app tracking transparency (ATT) policy, which mandates that users of iPhones must consent to businesses collecting their personal information. The message is clear. Personalization can be achieved by businesses by focusing on their own web properties. However, it is far easier said than done to create tailored experiences. As a result, they're turning to AI, which increases the importance of having ready-to-use data.

Encryption Of Sensitive Information

The process of safeguarding company data and preventing its loss as a result of unwanted access is known as data security. Encryption and decryption, as well as modification and corruption of your data are all part of this protection. Keeping data safe also means ensuring that it can be accessed by anybody in the company who needs it. Compliance with data protection standards necessitates a high level of data security in some industries. Data security is critical for today's businesses, regardless of whether they are governed by regulations or have to meet compliance standards. It can have an influence on both the company's most valuable assets and the private information of its customers. Rather than being the product of a hostile attack, the majority of data breaches are the consequence of careless or inadvertent disclosure of personal information. There are several ways that an organization's personnel may accidentally or maliciously share or grant access to, lose, or mishandle valuable data. Data loss prevention (DLP) technologies and better access controls are two options for addressing this serious issue.

More and more companies are shifting data to the cloud so they can collaborate more easily. As a result, it becomes more difficult to regulate and prevent data loss when data is stored in the cloud Personal devices and unsecured networks are the primary means by which users access their data. Unauthorized persons can gain access to a file in a matter of seconds, whether by mistake or design.

Attackers frequently employ SQL injection (SQLi) to gain unauthorized access to databases, steal information, and perform unauthorized activities. Adding malicious code to a database query is how it works. SQL injection is the process of altering SQL code by inserting special characters into a user-supplied input. Rather than processing a request from a legitimate user, it begins processing malicious code instead. If an attacker gains administrative access to a database, they run the risk of exposing sensitive consumer data, intellectual property, or both.

Organizations must ensure that their endpoints, such as employee workstations, mobile devices, servers, and cloud systems, are adequately protected against malware. In order to combat new dangers such as file-less attacks and unknown zero-day malware, antivirus software is still the primary defense.

A more thorough approach to endpoint security is provided by endpoint protection platforms (EPP). A machine learning-based examination of anomalous behavior on the device is used to detect unknown assaults by combining antiviral and anomalous behavior analysis. Security teams can also use endpoint detection and response (EDR) capabilities to discover and investigate breaches on endpoints as they occur, and then respond by shutting down or reimaging affected endpoints.

Big data security is concerned with the methods and technologies employed to safeguard enormous amounts of data and the processes involved in their analysis. Financial logs, healthcare data, data lakes, archives, and business intelligence datasets are all examples of big data. Large-scale data leaks, losses, and exfiltration can be prevented by big data security.

Protecting The Confidentiality of Personal Information

When it comes to protecting one's personal information, data privacy generally refers to the individual's ability to control when, how, and to what degree their personal information is shared or disclosed. A person's name, location, contact information, or online or real-world conduct can all be considered personal information. Many internet users desire to limit or restrict the collecting of their personal data in the same way that they might want to exclude others from a private chat. Data privacy has become increasingly important as the Internet has become more widely used over the past few decades. Users' personal data is frequently collected and stored by websites, applications, and social media platforms in order to provide services. Users may not know that some programs and platforms may collect and use more personal information than they thought they would, leaving them with less privacy than they had anticipated. A breach of user privacy could occur if other apps and platforms do not adequately protect the data they collect. internet privacy issues that users encounter, such as Monitoring of internet activities is a common occurrence. Although most nations require websites to notify users of cookie usage, consumers may not be aware of the extent to which cookies are monitoring their actions. Control of data is lost (With so many online services in common use, individuals may not be aware of how their data is being shared beyond the websites with which they interact online, and they may not have a say over what happens to their data), Transparency is lacking (To use web applications, users often have to provide personal data like their name, email, phone number, or location; meanwhile, the privacy policies associated with those applications may be dense and difficult to understand), A person's social media posts may reveal more personal information than they know, making it easier than ever to find someone online. Social media networks, in addition, frequently collect more data than users realize. Fraud, security breaches, and data sales on the black market are all examples of cybercrime, in which criminals attempt to obtain user data for nefarious ends. To steal personal information, some hackers use phishing attempts to fool users into disclosing their credentials, while others attempt to infiltrate corporate networks and steal sensitive data.

The process of scrambling data into what appears to be random data is known as encryption. To decrypt information, only persons holding the encryption key are allowed. To ensure that only authorized parties have access to systems and data, access control must be implemented Data loss prevention (DLP) can be used in conjunction with access control to prevent sensitive data from leaving the network.

It is critical for ordinary users to adopt two-factor authentication since it makes it much more difficult for hackers to obtain access to their personal accounts. In addition to this, there are a number of other technologies that can help protect users' privacy and keep their data safe. Data privacy cannot be protected solely by technological means.

Software for Data Transfer

Research is the key to finding the best data migration software. You can get a good notion of what's out there in the market by perusing review sites like Capterra, G2 Crowd, etc. In order to successfully migrate your company's data, you must first understand the fundamental characteristics that will assist you in this process. The following are crucial aspects to keep in mind:

- Mapping Data with ease

It is critical to map data from the source to the destination in order to complete a successful data migration. A graphical user interface with drag-and-drop functionality is ideal for making these operations accessible to business users and data specialists alike.

- Powerful Capabilities for Integrating and Transforming Advanced Data

By extracting and profiling data from the source, as well as modifying it to conform to the destination's schema, a successful migration project can be successfully completed. Connectivity to a wide range of heterogeneous sources, ETL, and data quality are just some of the characteristics that a good data management solution will provide, as well as the ability to convert and reorganize data to meet specific delivery needs.

- Increasing Interconnection

You need a migration solution that can handle many data formats, such as JSON, JSON, and EDI. You should search for a solution that has pre-built connections for business applications (Microsoft CRM, Salesforce, etc.) and databases like IBM DB2 and Teradata if you need connectivity.

- Effortless Data Movement

As a result of the automation of workflows and job scheduling provided by the best data migration software, more data can be moved more quickly. Automating the transfer process offers additional advantages, such as shorter implementation time, reduced project costs, and increased ROI, especially if your project contains huge data sets.

Look into the Source Codes

You must first identify and classify the source systems and data before you begin the migration process. It's preferable to start by organizing information like client names and addresses based on the target model, regardless of whether you're working with data from a warehouse, a data center, or somewhere in between. While the source data can include a wide range of fields, it's possible that some of those fields are redundant or otherwise irrelevant to the destination system. Consequently, it is essential to identify what data is required and where it is stored. In addition, you must determine which data is redundant and should not be transferred. The source data that will be imported into the target system will be identified after this phase is completed. You'll also discover any data gaps and, if possible, take into account additional sources to meet your business requirements throughout this step. Ideally, you should be able to group the data. You'll be able to work on numerous projects at once.

Carry out the Transfer

The next stage is to carry out the migration after a thorough testing phase. It is usual practice to put an end to the source systems while the migration is in progress. You may require a zero-downtime migration approach if the source systems must remain active at all times. Synchronization is employed to record modifications made during the initial load in this manner. Synchronizing the source and target systems after first loading ensures data completeness.

New System Transition

Once the data has been migrated, a decision must be made on when to implement the new system. If you have to, you may have to get rid of the old one. During the execution stage, audit logs and records are created to ensure that all of the data has been correctly migrated. You'll need to make some decisions based on these findings about how users will be moved to the new system.

Ongoing Improvements should be managed and monitored

When migrating data there will be a constant improvement in data quality. Some data quality issues in the source data may need to be addressed if the source systems are not discontinued. In addition, you'll have to monitor the data quality of the migrated data to guarantee that the data management process is as efficient as possible.

7. Interconnection of various systems (old and new) the API notion

One of the first concepts in computers is integration, which has been there since the very beginning of software. Web-based APIs, in comparison, are a relative newcomer to the block. APIs have been around since roughly 2000, but they haven't become a primary technology until the recent five to 10 years. APIs have nonetheless grown in importance in such a short period of time. When it comes to API integration, businesses have numerous alternatives, but there is only one correct one: APIs.

In the old days, organizations had to manually code a point-to-point connection between two computers or pieces of software. With today's technology, it's clear to see why this method would fail today: computers were still a luxury and software was in its infancy. One such link can take a long time to set up. In the modern workplace, there are hundreds of devices and apps, and it would take decades to link them all under the previous model. It's ideal for all the developers who have nothing else to do but sit around. Enterprise service buses (ESBs) and middleware have been popular as a solution to the integration problem in the last few decades. As a result, they serve as a single point of contact for all business IT systems. The service bus can be used instead of bridging solutions one at a time, and organizations can then let it manage the integration. It was an easy fix. That's not quite right. Although ESBs and middleware are capable, they aren't designed for use outside of the company's firewall. It's becoming increasingly difficult for middleware to keep pace with the proliferation of cloud apps and as-a-service offerings.

What does API Integration mean?

Application Programming Interface (API) is a term that refers to software that allows two or more applications to communicate data (messages) with one another. Think of an API as a virtual interface, similar to a touch screen, via which data can be entered, read or transmitted. Internet access is required to use web-based APIs.
API integration is the process of linking two systems together using APIs so that data may be exchanged between them. In

order to access a system from afar, APIs are developed to allow you to connect several systems together.

If two or more systems have APIs, they may communicate with each other in real-time, saving time and money, while also improving the accuracy and currency of the data. We used to be able to fax, email, or talk over the phone to get this information. It's all done digitally, without any human intervention, thanks to API integration. An API integration offers a new channel that allows our organizations to conduct business more efficiently and effectively.

API creation and management are critical tasks for any platform that can help companies make the most of their technology investments, whether on-premises, SaaS, or in the cloud and on either side of the firewall. APIs are essential for addressing the complex integration needs and concerns of today's modern enterprise. Businesses don't have to rip and replace their old systems or interfaces, but rather may connect all the essential data to make better business decisions and generate competitive difference by using an API integration platform." An API integration platform has the flexibility and scalability to shift to an all-cloud infrastructure when the time comes. It's hard to separate APIs from integration technology these days. It is possible to speed up the time to value of APIs by using integration technology. One coin has two faces: APIs and integration.

Each strategic integration infrastructure needs API-enabled technology, which are vital for API projects.

Why Do You Need?

More than half of firms struggle with integrating legacy programs, a problem that's only growing more profound as cloud-based applications expand. Many firms have invested considerably in their legacy systems, and they don't want to toss the baby out with the bathwater, yet they do want access to the cloud. Because of contemporary integration platforms, firms may speed up flawless end-to-end interoperability between their multi-entity ecosystem and their own in-house systems. Plus, by offering APIs to supplement your EDI onboarding processes, you may automate them and take on new ecosystem trading partners faster. Taking these kinds of IT modernization actions can help you improve your integration skills and, as a result, get started with B2B eCommerce more quickly and with greater confidence.
It's a well-known fact that time is money, and organizations who adhere to outmoded, manual processes in today's digital economy are doomed to fail financially. 66% of organizations are losing up to $500,000 per year owing to poor integration, while 10% are losing more than $1 million per year. Your integrations will go faster as a result of API

integration, which means that your money will move faster as well.

Any communication protocol can be unlocked with a modern integration platform at no additional expense. Such a platform utilizes established pre-configured connectors and templates, allowing you to set up fast and effective application integrations. A significant database can be linked to other internal systems via API integration to increase its overall usefulness to numerous departments. Customers and business partners can access your firm's APIs outside, allowing them to smoothly and instantly communicate mutually beneficial data with your organisation.

Data operations across your whole ecosystem may be more easily tracked and monitored, reducing the risk of exposure. Using next-generation integration tools, stakeholders may keep tabs on their most critical business interactions in real time via configurable dashboards. These tools also allow real-time monitoring and reporting. With an API integration platform, you'll be able to see if your new trading partnerships are in jeopardy. Because API integration allows existing applications to be kept while allowing them to communicate with other systems and applications in your B2B ecosystem, it is critical for logistics businesses, manufacturers, wholesale distributors, retailers, and others today. Making this a priority will improve your firm's processes, resulting in faster application integration and greater visibility throughout your whole end-to-end (internal and external) business processes.

Methods for Integrating API

Most companies' integration strategies now include APIs as a critical component for anything from applications and data to business ecosystems. APIs can be used in a plethora of ways to make integration easier for your company.

- Adaptation Programming Interfaces (APIs) for Product Administration and Monitoring

In the context of API integration, the term "headless administration" is used. It is possible to run a "headless" environment in which the computer does not have any kind of monitor, graphical user interface (GUI), keyboard, or mouse. In order to administer your cloud via an API, you must have access to an administrative GUI. You can handle the system "headless," which means you don't have to physically touch the keyboard to make changes. REST APIs now provide access to every aspect of data administration. Since the transformation is headless, the studio and runtime cannot be integrated, APIs are the only way to handle it effectively. This means that in many aspects, there are still certain holes that need to be filled.

Use an API instead of the GUI to perform activities like updating trading partners and AS2 connections or managing certificates. Instead, it's better to think of it as an API administration scenario that automates a number of important product functions, such as management, setup, and configuration.

- APIs for Uploading and Downloading Data

There are usually a number of secure communication protocols to choose from when it comes to moving data. There are a wide range of protocols that are used for file-based integration and include FTP, SFTP and AS2, as well as a secure portal for file transfers between people and systems. REST APIs can be used to upload and download files to and from the integration platform, as well as supporting APIs that can be set to upload and download files programmatically. When used in conjunction with file-based integration scenarios, these kinds of APIs can help a business work inside the traditional data transportation paradigm.

- Using APIs to Connect Different Systems Together

APIs given by outside systems are compared to those developed in-house. Core enterprise systems include Salesforce, which has a market share of about 20%, and NetSuite, which has remained a major player in the ERP market for many years. APIs are available from Salesforce and NetSuite in this situation and may be used for application-based cloud integration.

Integrate existing systems into one

An API is simply a method of transmitting data between a website, app, or software application and a user. In the same way that a waiter receives an order from a customer, relays it to the kitchen, receives the food from there and delivers it to the customer, an API receives instructions from a source such as an application, takes that request to a database, fetches the requested data or facilitates a set of actions and then returns a response to the source. The majority of states have systems in place that they have grown accustomed to, and making a wholesale change would require a tremendous amount of time and effort. When it comes to determining eligibility, caseworkers receive an application inside their benefits system, gather information from a variety of sources, and then manually enter that information into their systems to track determinations and re-certifications. With APIs, some of these external or manual tasks can be integrated without a total system overhaul or significant network investments. " A third-party verification API link, for example, can automate and integrate the agency platform. Back-and-forth manual data entry can be time-consuming and error-prone. In order to keep a social service application moving rapidly, the API can be connected and assigned tasks that order, fetch, bring back and offer up a necessary piece of information.

What does this signify for an agency's ability to get things done quickly? For instance, the health and human services department of a certain state recently merged with another. The state's system for determining eligibility for various welfare programs was integrated with a commercial income and employment verification solution in order to reduce the number of manual entries. The "swivel chair" of system entry and search was eliminated by using APIs to integrate systems. As a result, more than 3 million people receiving benefits received better service, as well as caseworkers who saved an average of five to ten minutes per case.

An example of a common API

In today's corporate world, APIs are an important tool since they let organisations grant other companies access to their resources while yet keeping their data safe and secure. Examples of common application programming interfaces that you might come across include the following:

Users can use their Facebook, Twitter, or Google accounts to sign in to websites, which is a famous example of a universal login API. An API from one of the most common services can be used to authenticate any user, saving them time and effort by not having to create a new profile for each website service or new membership.

Processing of payments by a third party: Using an API, for example, you can see the "Pay with PayPal" button on many e-commerce sites. Paying for things online can now be done without revealing any personal information or allowing access to unauthorised parties.

A travel booking comparison service aggregates thousands of flights and displays the most affordable options for every date and location. Access to the most up-to-date hotel and airline availability data is provided via APIs, which in turn enable this service. APIs greatly minimise the time and effort required to search for available flights or lodgings because of their autonomous flow of data and requests.

The Google Maps API is a nice example of an API in operation. There are a number of other APIs and features the app makes use of to give users with instructions or areas of interest in addition to the fundamental APIs that display static or interactive maps You can interface with the Maps API using geolocation and numerous data layers for planning trip routes or tracking goods on the move, such as a delivery van.

There are a few fundamental attributes for each Tweet: an author ID, a message (which includes a time stamp), and geolocation metadata. Developers can access public Tweets and responses via Twitter's API, and they can even post Tweets using the API.

Real time, current and efficient results can be achieved by using this tool

Companies can enhance their present technology while also assisting in the improvement of benefit judgments by using an API to connect current, automated employment and income information — frequently the building blocks of eligibility criteria. With the support of APIs, governments are able to more easily share the information needed to make benefit decisions while also ensuring that the integrity of the program is maintained, making it possible to help more people in need.

APIs: The Good, the Bad, and the Dubious

APIs, it goes without saying, have a lot to offer and benefit society as a whole. As a business owner or a customer, you've likely taken advantage of APIs and enjoyed the benefits they bring.
Businesses benefit from their increased effectiveness, increased reach, and increased output. An further benefit of

automation is that it reduces the amount of time and effort needed to accomplish a task. When it comes to ordinary activities, it makes life much easier for the average person. Smoother and more convenient operations can be achieved through the use of APIs, which focus on improving the client experience and providing personalized services. That's not all, either. APIs can even be put to good use in the pursuit of social good. API products can be used by government and non-profit organizations to provide services that benefit underprivileged and vulnerable populations. Adult education, safety, and other forms of support can be made available to individuals with the greatest need and the fewest resources by means of these programs. However, this takes us to the ethical issues raised by APIs and the concept of data sharing.

Hackers are more likely to get their hands on vital information when data is broken out of its silo and used and transmitted. Consumers are also concerned about the consequences of major organizations having access to their personal information. Even the government can misuse this information, which can be sold to other nefarious corporations.

Getting the Right Amount of Both

If privacy and data protection rules are followed, a wide-ranging business data programme looks to be the best alternative at this moment. Regulations from the government are important, but a data programme would allow consumers to understand what data is being gathered and stored as well as its intended use. The terms and conditions of the agreement would spell out exactly what the company can and cannot do with the data they collect from their customers. Consent is emphasised in these policies, ensuring that customers understand what the business and/or third-party organisations intend to do with their information. All businesses should have a data strategy that is consistent with their core values and overall vision. Furthermore, a company must be aware of the dangers of handling huge volumes of personal data and have safeguards in place to protect against data leaks and hacks. The trust, transparency, and goodwill fostered by a strong data program goes a long way in dealing with sensitive information like personal data. A business' growth is aided by ethical guardrails, not hindered by them, because of this. Customers who trust you with their personal information will be more willing to do business with you again if you can guarantee their privacy and security.

API Integration Has Many Advantages

API integration has numerous advantages for your organisation as a whole, but let's go a little further and see what else it can do.

Earlier, we spoke of more fruitful interpersonal interactions. There's arguably nothing more critical than providing a level of service to your customers and partners that is in line with their expectations these days. Indeed, customer experience is more important than income since only by making it easy for customers to do business can you ensure that they'll keep coming back. Since they have so many options, their allegiance is typically short-lived. Interfacing two or more disparate systems has long been one of the most challenging and irritating aspects of IT. APIs facilitate integration by simplifying the establishment of connections, the transfer of data, and the transformation of that data. Your company operations will run more smoothly if you allow multiple systems or applications to interact and share information effortlessly.

In addition, with today's pre-built solutions, you may speed up the delivery of custom API integrations and save on development expenses, or enjoy having your developers build apps that are vital to your business rather than spend their time implementing sophisticated API-based integrations themselves. Manually doing chores takes a long time. Your life will be easier if you use API integration to automate routine operations. If we look at APIs from both a software development and a business cooperation perspective, their significance grows significantly. These resource exchange interfaces that may be read by machines are similar to delivery services that work behind the scenes to provide the necessary connectivity.

The age of digital technology is very new and the API economy is even younger. We need more time and research to see whether the use of APIs is ultimately good or bad for mankind. However, we can progress into the future by utilizing the good that APIs bring while setting boundaries to stop the unwanted side effects. A data program would be the first step to balancing APIs and ethics. Customers trust businesses when putting personal data into their hands, and protecting their information will benefit everyone in the equation.

8. Importance of training for business users and for the adoption of new system

Digitization is becoming more and more significant in the corporate sector, since it helps to drive growth and efficiency by utilizing digital processes. As a result, the company's infrastructure and culture are dramatically altered, requiring its people to learn new skills and change their mindsets to adapt to the new environment. You'll go a lot closer to accomplishing your digital transformation goals if you provide your employees with digital marketing training. In fact, according to Adobe's 2019 Digital Trends study, digital-first organizations are 64 percent more likely to meet or exceed their primary business objective. If you have the ability to adjust to these changes, your firm will thrive and expand even in the face of any setbacks.

Development of a New Technology Training Program Is Crucial

Every new technology project is met with a few dreams of user acceptance in the imagination of an IT manager. At times, we're acting in the best interests of the company's top brass. That technology was never a goal of ours and we are confident that no one in the organisation will make use of it. Other times, we take care of the implementation. End users will be on their knees thanking us for providing them this new application, as we have visions of all the amazing ways it will benefit our organisation. New technology may be required because the current system is no longer adequate. However, the old system is beloved by all end users, who are blissfully unaware of its security, reliability, and resource consumption. This is how we believe they'll break into our office to tar and feather us: with their torches, with their pitchforks. If the old system is removed, there will be riots. When it comes to a new technological project, our assumptions of how people will respond to it are unrealistic — some will like it, some will hate it, most will be apathetic, and all will use what is most convenient for them. Most people have this reaction when a new technology comes online. This means that most of the time, the system will go unused. There will be a small percentage of employees who use all of the system's features, while the great majority will get familiar with the basic functions and never bother to learn about the more advanced ones that could help them do their jobs better. However, if we as IT managers are proactive in producing training materials and pushing our people to learn the system, we have the best chance of success.

Avenues to take a lot

An instructor explains the basics of the new technology to a group of students. For the most part, they're long and tedious. Many of your coworkers aren't going to take this seriously. Look for methods to make it more enjoyable by bringing some variety into the routine. Make it a contest to see who can utilize an app the fastest. It may be possible to spread out the sessions across the course of a month or many months.

The use of video training, if available, can significantly aid in long-term adoption. If video tutorials are easily available, your employees won't need to seek you for help. These movies should be broken down into relevant applications so that they aren't too long and employees can quickly find what they need.

Similar to video instruction, but without the visual component. Employees can go back and review lessons they've already learned thanks to online training. You don't have to hold their hands every time they use technology if they have a cheat sheet available online.

Don't be hesitant to bring in a design team and write instructions for the new technology, if you have enough resources. You could print a bunch of employee handbooks and distribute them to each one. Write them so that even non-technical staff can understand them. Make it easier for them to find their way around by providing visual aids. Using the system doesn't necessitate that they fully comprehend its inner workings, thus don't go into too much detail.

Though time-consuming, certain new technology makes personal training unavoidable. To encourage participation, schedule personal training sessions for employees at predetermined times throughout the week. The only way to ensure that you pass your math exam is to hire a tutor. The same reasoning can be applied.

You can also offer a combination of these training methods. It's possible to start with a comprehensive instructor-led training session and then provide on-demand videos of new technological training programs to help employees keep current. People who are in need of personal training can get it from you. You can design handouts to distribute to students following classes with an instructor.

Actually, you'll know exactly what to perform in terms of training based on factors such as your own internal resources as well as the individuals you'll be instructing. Regardless, while the build is underway, resources should be developed in order to begin teaching as soon as the technology has been fully implemented.

Knowledge of Digital Marketing Can Help You Improve Your Customer Experience

A successful digital strategy may help your company's digital transition by optimizing crucial parts of your business—specifically the customer journey. Visibility and revenue can only be increased by implementing effective marketing methods. Your company can increase its reach and better target clients who desire or need your items by switching to digital marketing tactics. It has been estimated that digital transformations focused on customer experience can enhance customer satisfaction by 20-30% and provide economic gains of 20% to 50%. Your staff may help you achieve these objectives by learning how to execute successful ad campaigns, establish large-scale digital marketing plans, and work with analytics to improve results through their training. Digital transformation might be helped along by your company's marketing department.

Training in digital marketing teaches you how to make the most of your digital channel

One of the numerous advantages of corporate digital marketing training is the ability to fine-tune and improve your digital channels. Marketers are now able to obtain better outcomes with less effort because to the innovations made possible by digital marketing. To mention just a few, there's social media marketing, SEO, and marketing automation in here. Digital marketing training involves giving your staff the tools they need to succeed. Your staff may learn how to best manage and optimize various elements of your organisation, including your website, social media accounts, and analytics and automation technologies. Customers and leads will respond to personalized messages created by your personnel with these new digital marketing talents. Your company's growth will also be boosted.

Investing in the digital skills of your employees improves both communication and productivity

While working on real projects, employees with corporate training can improve their communication and efficiency through teamwork, gaining significant experience. A multi-channel marketing model, where leads and consumers are targeted via a variety of different marketing platforms, is a better fit for your personnel with these skills. Your sales, marketing, and customer service staff can all benefit from the digital marketing abilities they've learned during their training. An effective and sustainable digital transformation is the consequence, affecting your company's infrastructure as well as its performance.

Definition of user adoption

The process of getting new consumers to start using your product and stick with it over time is called "user adoption." A new method or product that is more suited to their needs and more effective in helping them reach their goals has replaced the previous one. There are two main areas to focus on in order to increase user adoption:

Onboarding a new customer: Your onboarding process aims to get new customers up and running with your product as quickly as feasible, so that they may begin reaping the benefits as soon as possible.

Management of change: In order to retain new customers, you need to make your product better than their previous product or system (regardless of whether they utilized a competitor's product, spreadsheet, or a variety of manual processes).

What is the significance of user adoption

A successful software implementation begins with identifying the need for new software, conducting market research, determining a solution, and drafting an implementation plan. As a result, one of the most critical measures in the success of any corporate software project has to do with whether the program is being used. The new system's advantages aren't fully realized if the software isn't used to its full potential or if users are unwilling to give up their old systems. The implementation of software should not be limited to an IT project with the purpose of technical deployment, but rather should be a process that involves the entire organisation. Management and users must be included in the process as well. As a result, the new system will be widely adopted, and the software's worth will be maximized. If the new program can reach consumers faster and provide them with the resources and support, they need to use it, it will be more successful.

Understanding how the product will be used is critical

Deployment is more than just an academic exercise. In order to ensure that the software is suitable for its intended usage, it is vital to first identify and address user pain points and difficulties. Considering moving your activities to the cloud because of data synchronization issues? When it comes to your scheduling and financial tools, do customers have any complaints regarding the lack of integration? To guarantee that any new solution addresses existing challenges and adds value to end users, it is important to know how consumers expect the system to operate in the future. Once you know what your customers want and how you can help them achieve it, the rest is a piece of cake.

The foundation of a user adoption strategy is training

In order to get people to utilize new software, you must first identify the problems they are trying to solve, and then devise a training strategy that addresses those issues. That way, you can answer the end-user query about the value the new software provides, and in doing so, automatically provide a justification and motivation for its adoption. A user handbook by itself is not an interesting motivator for behavior change, even if it does contain all the necessary technical instructions. However, a convincing demonstration of the software's advantages might act as a powerful motivation.

Maintain a diverse training program

You may begin to build your training strategy once you have a clear understanding of the benefits the implementation will bring to your users. There are a variety of ways in which people learn, and you want to ensure that all of your users have a chance to become acquainted with the new system. When it comes down to it, your training strategy is nothing more than a communications strategy, and as such, it must incorporate as many different types of training as feasible. To normalize the concept as much as feasible, hold presentations, provide hands-on experiences, and devise a corporate communications strategy. There is still a role for detailed instruction manuals, but they aren't the most effective way to win over users' hearts and minds.
A training session isn't a one-way street. Aside from increasing user acceptance, user training can also assist you discover snags that may have gone unnoticed had you not had the feedback of actual users.

Training pays off in the long run

As daunting as it may seem at first, running a complete user training program in conjunction with technical implementation of a new corporate software system is well worth the time and money invested. Users should be thrilled about the new system and eager to get started using it on implementation day in an ideal world. The most important thing you can do to get to this position is to obtain some training. Adoption is significantly easier when consumers are familiarized with the essential concepts through proper training.

9. Designing a solution and migrating to the cloud

Finding solutions is a way of life in the corporate world. The goal of companies is to help their customers, but they must first overcome their own issues in order to do so. Solutions aren't always easy to find, however. However, they are rarely all three at once. As a result, they must be well-understood by everyone in the company. If this isn't the case, it will be difficult to get everyone on board with a single objective. All solutions must be created in a way that is accessible to everyone and maximizes the pace, cost and quality of development. We call these visual representations of business solutions, solution designs, the result of the process of identifying solutions. High-level graphic representation of business solutions that define how the corporation will minimize implementation time and cost and maximize solution quality are called "solution designs."
Keeping up with market trends and the competition necessitates that IT workers always explore for methods to increase operational efficiency and take advantage of the

insights that new technology may offer. The previous paradigm of custom solution design necessitates expensive developer skills and tedious periodic updates, making it a popular choice for technology leaders. Because of their low-code environment and ability to swiftly roll out apps, these new platforms allow for increased user productivity and overall usability.

Development in a coding language that does not require a lot of

An application's user interface, user experience, functionality, and technical procedures are all part of the solution design process. Advances in solution design can include apps that combine data from several corporate systems or integrate cloud connectivity, the latter of which is gaining major acceptance globally and essential for features like digital twin and IIoT connectivity to be unlocked. Patching together "good enough" solutions and adding customization to obtain functionality is a recurring topic in the enterprise resource planning (ERP) area that often renders systems inefficient and unreachable, rather than this. The advantages of low-code application development as a means of custom solution design are being considered by leading enterprises. Using a visual IDE and a modular

workflow that includes drag-and-drop capabilities, developers of all skill levels can benefit from low-intuitive code's interface while also speeding up development.
With low-code development you have more control over coding boundaries, a more cost-effective environment that enhances developer efficiency, and application "blocks" which are constantly updated by the provider for security patches or other technological advancements over time compared to traditional programming methods.
You'll need the ability to quickly roll out scalable applications when embarking on a digital transformation strategy, which will necessitate customizing applications, altering the fundamental infrastructure, optimizing current systems and removing non-core functions like product information management, warehouse management, work-in-progress (WIP) inventory control, customer service operations and supply chain operations from your existing systems.

A Quicker Changeover

If you work in a firm that relies on technology in any way, you know how quickly new solutions and IT services become obsolete. If you want to be successful in today's highly competitive marketplaces, where ongoing transformation and innovation are a need, your organisation must be dynamic, too. It is possible to breathe new life into your development team and processes with a low-code platform that removes complexities when building new applications, allowing your organisation to change at the speed of thought and innovation rather than at the speed of in-house bureaucracy and backlogs of developer projects.

Interactiveness

Working smarter rather than harder may be applied to every business, especially those that frequently have to launch new projects and applications in a short period of time. Because of the block-driven, modular design of a low-code platform that allows your organisation to accomplish unique solution design, your staff will be able to develop more apps in less time, increasing productivity across the entire IT team.

Cost Reduction

Regardless of sector or size, companies are continuously seeking for methods to improve their capital management processes. Using low-code development as a cost-saving method is a good fit for your firm because:

Decreases in overall human expenses can be achieved by developing more apps in less time.

Because there are no external developers needed to build applications, variable costs are eliminated.

Because vendors are responsible for the upkeep of low-code platforms, your team does not have to devote resources to rewriting code or implementing security updates.

While still adhering to your company's particular business processes, custom applications can take the role of licensed and frequently redundant solutions.

Advanced Personnel Skills

Businesses have traditionally adopted the waterfall process, in which each phase must be accomplished before the next can proceed. An increasing number of firms are adopting agile development methods, which allow them to make iterative adjustments during a project by distributing talent across teams based on the capabilities they need. More rapid app development will be possible for employees of organizations with improved orchestration, standardization, and automation. Unifying application lifecycle management by bridging the gap between development and operations teams.

What is cloud migration, and how does it differ from traditional migration?

Moving digital assets, services, databases, IT resources and applications to the cloud can be done in part or in full. Moving from one cloud to another is also part of cloud migration.

A growing number of businesses are going to the cloud to reap the benefits of cloud computing instead of clinging to antiquated and increasingly inefficient legacy infrastructures like aged servers or potentially unstable firewall appliances. As a result, a large number of businesses are making at least a partial move to the public cloud. For real-time and updated performance and efficiency, cloud migration is essential. As a result, careful research, planning, and implementation are necessary to ensure that the cloud solution meets your company needs. In the process of deciding how to migrate your data to the cloud, it is crucial to remember that it is not just about getting there.

Process of moving to the cloud

It's critical that businesses fully comprehend the ramifications of any decisions they make.

It's smart to start by looking at the full cloud experience from beginning to end because it can be a transformative move for your entire business. With this information, it will be possible to identify which skills and activities are required to carry out the three key stages of cloud migration efficiently. To begin, answer the question, "What business value will a move to the cloud provide? "; then, develop your strategy and financial case. In the end, moving to the cloud is much more than just a technological experimentation. What the company is trying to accomplish needs to be the driving force behind it.

A cloud migration strategy and business case can be developed around these goals.

- Determining whether apps will be migrated to the cloud, and if so, to whatever type of cloud environment, is an important part of this approach.

Apps that have varying loads, are public-facing with worldwide reach, or will be modernized in the near future are ideal candidates for the cloud. These include those that are too dangerous or difficult to relocate, as well as those that don't offer a good return on investment. It is critical to determine this from the beginning of the migration process to ensure a successful outcome.

- Risk management is an important part of any company. It is important to keep in mind how your applications might operate as a result of large infrastructure changes, even when businesses demand increased flexibility, cost, and control.

If you don't know where you are now, you can't know where you're going to be in the future.

- The Cloud Migration Assessment helps transfer planners to make educated decisions, minimizing risk while ensuring service level agreements are maintained following cloud migration, thanks to application identification, dependency mapping, and risk assessments based on actual usage.

When it comes to cloud migration, the real work begins—the actual transfer of data to the cloud.

Typically, this involves updating existing applications for the cloud, developing new cloud-native apps, and modifying the architecture and infrastructure of the current system.

The ultimate goal is to develop a new operational model and culture for technology that will allow the business to innovate more quickly, effectively, and efficiently.

A successful move necessitates the use of automated management and migration solutions. Additionally, they aid in expediting the migration and delivering consistent and repeatable results. As a part of the cloud migration factory, they can help speed up the process even further when combined with specialist expertise and solution accelerators

Cloud journey management plans are also crucial at this phase to keep the effort on track.

The advantages of moving to the cloud

The cloud can have a huge influence on firms who migrate to the cloud.
Reduced total cost of ownership (TCO), shorter delivery times, and more room for innovation are all benefits of this. The ability to respond quickly and easily to shifting customer and market needs is a key benefit of using the cloud.
To keep up with the rise in online demand and the spread of remote working, businesses have been moving their services and data to the cloud in recent months. A cloud transformation is being accelerated by firms that have already made the transition to cloud computing. Others may ask, "Why did we wait so long?"
There are numerous advantages of moving to the cloud: increased agility and flexibility; greater ability to innovate; easing of resource demands; improved customer service; reduced expenses; rapid business results; a shift to everything as a service; and simplified IT. Better control of usage, scalability in the cloud, and improved performance are all benefits.

Security

It is possible to make the cloud more secure than traditional network systems if it is implemented correctly. In this piece, we'll go over five advantages of moving to the cloud. Systems, networks, and applications in the cloud must be designed and maintained in accordance with the "shared responsibility" paradigm in order to ensure the cloud's security. To ensure cloud security, you must take responsibility for your own part in the process. A primary consideration for cloud service providers like Amazon is security as part of the "shared responsibility" they have undertaken. Unlike other networks, Amazon's cloud service was built with privacy and security in mind, unlike other providers. According to a Microsoft Office 365 survey, nearly 94 percent of small and medium-sized businesses (SMBs) value the cloud's greater security. Your data can be accessed no matter what happens to your physical equipment, because it is stored in the Cloud.

The cloud provides considerably better security than traditional data centers since it centralizes all of your business information and data. Additionally, many notable clouds service providers include a variety of built-in security capabilities, such as security analytics, regular upgrades, and cross-enterprise visibility, in their offerings. When it comes to protecting your business data and applications, many cloud service providers handle some of the more difficult security issues, such as preventing unwanted traffic from accessing the machines on which your business data and applications reside and ensuring that automatic security updates are applied. There is no need to worry about the security of your data in the cloud if your company has special compliance requirements or concerns.

Scalability

By moving to the cloud, your firm will be able to grow or shrink more easily in accordance with its IT needs and strategic goals. It can be tough for organizations to keep up with changing client expectations if they have an IT solution that is no longer relevant in a few years. It is now possible for companies to drastically alter their infrastructure and workloads to meet the demands of the present without being tied to outdated equipment and assets. Control over your resources is now at your fingertips with cloud computing. With other options, such as contracts, minimum terms, and one-size-fits-all plans, this is impossible.

The money you save by not having to rely on expensive data centers that rarely meet their promises and potential is used to grow your business. With a cloud transfer, your company may grow and expand without disrupting its current infrastructure. Your business and customer experience will not be affected by the growth of applications and data. With a fraction of the staff, your company may benefit from the lower overhead expenses and higher agility that the cloud offers to boost efficiency, productivity, and customer happiness. This is because you no longer have to lease pricey locations or invest in expensive technologies and equipment that take up important time and space. As your company expands and changes, you may be able to free up this space and capital through a cloud migration. This is the most significant advantage of moving to the cloud for many businesses. To ensure a successful migration to the cloud, work with consultants who have extensive experience in this area.

Price Factor

You can cut operational costs while also enhancing IT operations by migrating to the cloud. When your data is hosted on the cloud, you only pay for what you need, and you don't have to maintain expensive data centers. With cloud technology, 82% of small and medium-sized businesses (SMEs) report lower operating costs, while 70% of these businesses are reinvested with the savings, according to a

Microsoft Office 365 survey. Businesses of all sizes can benefit from reduced operational expenses and increased IT efficiency and throughput through a move to the cloud. According to cloud solutions provider VMWare, major enterprises can save 40% to 50% of their typical IT running costs by shifting their in-house data centers to a public cloud service provider. Cloud computing's cost advantages are innumerable.

In addition to the hardware needed for web servers, the public cloud host provides all of the services needed to maintain and upgrade those servers, including security, maintenance, and stack configuration. Pay-as-you-go models are common in the public cloud. In addition to the upfront investment expenditures, the total cost of a data centre includes operational costs as well. Additionally, you'll be responsible for continuing support and maintenance as well as the costs of electricity, cooling and staffing. More and more CIOs are going to the cloud in order to boost flexibility and alleviate strain on capital budgets. Deloitte estimates that 62% of yearly IT budgets are spent on internal maintenance in commercial and professional service firms. Migration to the cloud, like anything else, allows your company to take advantage of economies of scale. As a result, public cloud providers such as AWS, Microsoft, IBM, Google, and Oracle are able to minimize the costs of their maintenance, upkeep, electricity, cooling, and staffing per server unit compared to a data center maintained by an independent company. The easiest way to save money when moving to the cloud is to follow these cloud migration best practices.

Integration

Moving your business's systems to the cloud enables you to connect them effortlessly and increase your services' efficiency. Data centers, like any other machine, are subject to wear and tear over time due to growing demand and diminishing efficiency. To maintain its data center software and hardware up-to-date and running properly, businesses have traditionally had to undertake infrastructure refreshes on a regular basis. Costly and time-consuming, this work is a constant drain on time and resources. The end of a hardware refresh cycle can now be used as an opportunity to move apps to the cloud. Hardware and software updates are handled by the cloud provider in a cloud environment, which saves money, time, and ensures that applications are always supported by the most current infrastructure. Based on the demands of both businesses and consumers, cloud services and applications are continually being updated and expanded upon. This implies that your new cloud environment may grow and extend to match your business needs while allowing your team to do more than ever before. When you move your business-critical applications to the cloud, they will be able to respond dynamically to variations in traffic. To put it another way, your organisation can boost productivity and profitability by optimizing to scale up or down to meet demand and only employ the resources you require. When you use a cloud service, you don't have to worry about the complexities of your infrastructure. There is no need to worry about your employees' productivity because of the ease and convenience of cloud remote access.

Access

It doesn't matter what happens to your hardware because all of your data is kept in the cloud. As an added benefit of moving to the cloud, your employees will be able to use any device, from anywhere in the globe, to access the data they need for their jobs. With so many possibilities for growth and expansion while still satisfying operational requirements, this is a game-changer for your company. Instead of wasting time and money due to unproductive snow days, your staff can now work productively and safely from any location. Monitoring for mission-critical apps and equipment is available from several cloud services. When an outage or possible issue with an app is discovered, it can drastically shorten the time it takes to restore services and fix the problem. With this method, you'll save both time and money by not having to keep track of your services yourself. In the event of a catastrophe recovery after an outage, you'll want to have good backup and logging systems in place so you can figure out what went wrong. Backups and logs can help you figure out what went wrong in the first place, and you can use them to restore your system to its previous state. To deploy, update, or fix issues with any of the machines in your cloud environment, you and your team will no longer be confined to a single place. In comparison to the conventional problems of an on-premises arrangement, this provides a more flexible solution. Because everyone on your team will be following the same set of procedures for provisioning and deploying resources in the cloud, collaboration will be much easier. The need for openness and adaptability has never been greater for businesses of all sizes. A job in 2020 means being able to work from any location where your clients require you to do so.

10. Obligatory Role of Ethics

As customer expectations continue to rise, so does the requirement for enterprises to adapt to market dynamics with ever-increasing levels of business agility through digital transformation.

With cloud computing, robotics, artificial intelligence (AI), and big data paired with streamlined operational models, firms can drive innovation and quickly adapt to internal or external events at lower costs than ever before.

Humans are as crucial to digital transformation as computers, and these findings should help us keep this in mind. In order to achieve long-term success in a digital environment, organizations will need to guarantee that they recognize the need of acting ethically above all other considerations.

When it comes to ethics, it's not about what people do, but rather how they should act in certain situations. Since it was previously difficult for the subject to obtain acceptability in the business sector, things are now much better.

Ethics is not only vital to many firms, but is now a key difference in a highly competitive market where reputation and values are as important as the products and services. Many organizations now consider ethics as an important differentiator.

All firms that are undergoing digital transformation should take ethics very seriously, but the most difficult issue will be to get people to take responsibility for their own actions. When it comes to morality, there is no longer just one correct response, but rather a variety of possible outcomes depending on one's perspective. It is not uncommon for people to be faced with ethical dilemmas in which they must choose between two or more unpalatable options, each of which has implications for other parties that must be taken into account.

From an ethical standpoint, digital workers and executives will have to figure out what the "correct" thing to do is. It doesn't matter how much ethics training or knowledge there is in an organisation; what is ethical can be interpreted in a variety of ways by people from many walks of life, and in a global, fast-paced digital society, this leaves a lot of opportunity for interpretation.

It is possible for individuals to make actions that could have negative implications for others even when the correct course of action is evident. Ultimately, being ethical means having the ability and moral bravery to defy established standards and act ethically. What does it mean to be ethical in the digital age, and how can we apply our ethical principles in the real world?

Digital businesses must adhere to business ethics

Integrity and trustworthiness will be just as critical as the technological issues that application integration, cybersecurity, and data governance confront in the future, for example. For example, the individual's rights must be honored by embracing the need to avoid harming oneself. As a result of unethical behavior, companies are likely to receive negative attention from local, national, and even international media outlets.

There is no way that the events involving Volkswagen, in particular, will be forgotten. As a pioneer in automotive innovation, Volkswagen was formerly well-known throughout the world. However, the company's unethical practices are now more likely to be remembered than its earlier accomplishments.

Despite the potential for long-term benefits from digital projects, they must be weighed against the challenges created by dwindling confidence due to concerns about how some organizations are utilizing digital technology.

Concerns about data security and confidentiality persist; not only are breaches occurring at an intolerable pace, but corporations have also been unwilling to alert people affected when they do. Existing accountability procedures may no longer be viable or even undermined by the rapid evolution of AI and machine learning algorithms to make autonomous choices.

Trust in the digital economy is only possible if corporate executives are willing to make ethical decisions despite working in highly competitive circumstances where the temptation to swiftly deploy digital services will win over the need to evaluate the ethical consequences of doing so.

Design for confidentiality, integrity, and safety

Despite the potential for innovative and lucrative insights to be gleaned from big data, there are also hazards associated with what may be considered as immoral activity, whether deliberate or not. It's not just a legal, it's an ethical one how data is collected, maintained, and used.

Transparency and trustworthiness are essential in today's digital environment. This means that organizations must not misuse data in ways that are invasive, manipulative, or insulting to others.

Being transparent means that firms must disclose their goals for data usage and allow their customers to offer their approval. The amount of client data that organizations acquire and sell is frequently questioned. This problem stems from the fact that customers and service providers don't have a clear understanding of the value exchange. There should be no monetary gain from gathering personal data unless it is shared equally by the parties involved.

An important premise in ethics is the concept of "informed consent," which refers to permission given with full knowledge of the possible repercussions. Organizations collecting data, especially personal data, face a substantial difficulty because the intended use of the data is often unknown at the time of collection.

Even though it may be impossible or prohibitively expensive to get informed consent, the principle should still be applied, when possible, in the design and development of digital services. If agreements are required to use digital services, informed consent will still be a topic of controversy because organizations will need to assess both their validity and extent of consent.

Facial recognition applications are being used in certain countries to collect personal data for reasons that are likely to be less than sincere, and concerns have already been raised about the ability of mobile phone software to track individuals' movements even when the location services feature on the phone is inactive.

Inspire faith in others

Customers of digital services and data must be able to rely on them, regardless of whether they are individuals, groups or companies.

In order to be useful to customers, those that collect and manage data must adhere to the premise that its integrity must be ensured. An organization's obligation to protect data integrity necessitates rigorous governance and audit procedures for the data it has. Companies must be confident that the information they've provided is accurate.

Digital infrastructures allow data to be made available to others for a variety of purposes, including validation, replication, and analysis, in addition to simply storing it. Consumers must have confidence in the data driving their services if it has a clear origin, full traceability from source to user interface, and sufficient quality for its intended use.

A considerable amount of risk is imposed on individuals who consume the data if the provenance and accuracy of the data cannot be checked. Once processed, any actions taken as a result cannot be reversed.

There were only 11% of Americans who felt 'somewhat confident' in the privacy of their personal data on video and social media sites surveyed by the Pew Research Center in 2014. The difficulty of segmenting customers based on their privacy attitudes, which are context-specific and resist generalization, is one aspect of the dilemma. In the UK and the US, nine out of ten internet users would reject doing business with organizations that don't respect their privacy, which is a problem for businesses.

Without acknowledging the essential principles of privacy, security, and integrity, organizations that are racing to produce 'the art of the possible' are likely to face high levels of ethical risk.

Be wary of preconceived notions

A number of factors might contribute to unintentional ethical behavior, but one of the most common is the presence of unconscious biases.

When people seek or interpret information in a way that confirms their ideas, hypotheses, or expectations, they disregard information and opinions that challenge them. This is known as confirmation bias. An ethical concern for people who want research and testing to be conducted with the utmost integrity and objectivity is that confirmation bias could lead to the 'cherry-picking' of data for analysis, study, or testing.

It is likely that consultants will face one of the most difficult problems during their engagements when it comes to digital transformation programs, which are particularly vulnerable to bias. Because the line between expressing commitment to stakeholders and acting in a way that offers them an unfair benefit is not always apparent.

A team of consultants on behalf of their client has the right to ensure that the tendering process is fair and transparent for companies that respond to it. Competitors are entitled to demand a level playing field in the bidding process, which they have invested time and money in. In the future, if this right has been infringed, it could have a detrimental impact on the way they work with other organizations.

Ensure that there is accountability

Digital services with artificial intelligence and machine learning capabilities are increasingly incorporating inference models and algorithms.

Concerns have emerged concerning the ability to maintain clear accountability frameworks due to their potential to mix social data with decision-making engines. As an example, if financial services businesses utilize algorithms to make judgments that would typically be made by qualified and regulated personnel, concerns need to arise about where accountability resides.

Digital service providers must consequently make sure that these services are not exploited to avoid or diminish corporate accountability

Promoting a more ethical way of life

One way to think about organizational culture is to think about the values, beliefs and conventions that guide the behavior of its members.

The values of a business, such as 'offering an outstanding digital experience' or 'excellence through innovation and teamwork,' are particularly significant since they express what the organisation stands for.

A list of corporate principles can be found on the company's website, but that doesn't mean you've adopted them. Organizations that express their values and utilize them as a guide even if it means making difficult decisions are value-driven. Ben and Jerry's, which began by selling scoops of homemade ice cream at a loss and has now created a company known for caring more about people than money, is a good example of how such principles are feasible.

With the promise of opportunity, value, and success at every level of the business, digital transformation must be delivered with justice, honesty, and integrity. As a result, digital businesses have a moral obligation to protect their employees from taking risks that could have negative consequences for both the employee and the business. The importance of a person's professional reputation should not be overlooked.

Digital Success

The successful organisation in the digital economy will be the one that actively employs ethical values such as trust, honesty, fairness, secrecy, and accountability to do the 'right' thing and make judgments that are above reproach. The degree and type of success that a company achieves will ultimately be determined by how successfully its employees, suppliers, and other important stakeholders adopt and share these values. To avoid creating more rules that the organisation must follow, digital ethics are always aligned with business activities at their heart. Retroactive control will be replaced by the integration of digital ethics concepts in the future.

Successful digital transformations are distinguished by broad support for and confidence in the use of digital technology by its stakeholders. This can only be accomplished if organizations put a high priority on digital ethics early in the transformation process and integrate it into all aspects of their business operations. Here, the emphasis is on their willingness to accept personal responsibility for adhering to their own set of moral standards and culture.

Ethical Career Development

A sense of moral humility leads people to see temptations, rationalizations, and circumstances as just as likely to drive them as the best of us as the worst of us to act unethically. Seeing it as a long-term goal helps students perceive character development as a worthwhile endeavor.

People are often aware of what they should do in the future but tend to focus on what they want to accomplish in the now. Preparing for ethical challenges is vital. An ethical mirage is the tendency to exaggerate our own virtue in the future.

The first step in overcoming this prejudice is to gain a clear picture of your own strengths and flaws.

Setting goals can also serve as a foundation for ethical conduct. In both their professional and personal lives, professionals routinely set goals, yet few do so when it comes to ethics. However, even the most meticulously crafted objectives are nothing more than well-intentioned wishes. As a result, they need to be bolstered by personal safeguards—that is, behaviors and dispositions that have been proven to bring out people's best selves.

The moment you've decided to lead an ethical life, don't be afraid to tell others. Even while no one like a self-righteous attitude, subtle moral signaling can be beneficial, especially when it's directed at coworkers. Openly discussing moral dilemmas and how you would respond is one approach to do this. You may also create a reputation for doing the right thing.

Think ethically about technology

When it comes to technology disruptors, ethical decision-making is becoming more and more critical for tech-savvy CEOs and their enterprises. Using disruption-oriented terminology and phrasing isn't enough. Disruptors can also pose ethical challenges for organisations, so they need to learn to identify them and develop the muscle memory to choose the best course of action. This muscle memory can be developed by leaders and people who are committed to ethical decision-making and who foster an environment that encourages it.

An inclusive, cross-functional responsibility should be pursued. Ethical considerations should guide all of your actions from the beginning. Take a holistic, tech-savvy approach to ethical technology. Make it specific, relevant, and adaptable. Verify that it goes beyond just compliance. Give your employees the tools they need to take action. Make sure your strategy is flexible enough to change with the times. No matter how technologically proficient your company is, it is becoming a technology company. This is why ethical technology is important. Firms with a high level of digital sophistication place a higher value on ethical technology, but it's possible that this trend has accelerated as a result of the increasing attention being paid to ethical issues in the media and political debates. In order to ensure that the judgments they make on how to deploy disruptive technologies are not only strategic but intelligent, leaders should assess technological options from different aspects. The time is now for leaders who have yet to make ethical technology a priority to take advantage of this opportunity. Keep in mind that the rapid pace of market change and technological advancement may make even the most recent ethical technology policies ineffective in the face of today's threats.

The future of ethical technology lies in the hands of leaders who will make it a priority, instill it in the company's culture, and implement ethical decision-making processes based on technological expertise and a wide range of input. An ethical technology perspective can help firms anticipate and respond to ethical issues that may arise in the future.

www.ingramcontent.com/pod-product-compliance
Lightning Source LLC
Chambersburg PA
CBHW082107220526
45472CB00009B/2078